To William Shakespeare for his immortal words *"for I have dream'd of bloody turbulence, and this whole night hath nothing been but shapes and forms . . . "* and to those who have perpetuated this dream, this volume is dedicated.

Turbulence in Mixing Operations

Theory and Application to Mixing and Reaction

Academic Press Rapid Manuscript Reproduction

Turbulence in Mixing Operations

Theory and Application to Mixing and Reaction

edited by

Robert S. Brodkey

Department of Chemical Engineering
The Ohio State University
Columbus, Ohio

Academic Press, Inc. New York San Francisco London 1975

A Subsidiary of Harcourt Brace Jovanovich, Publishers

ACADEMIC PRESS, INC.
111 Fifth Avenue, New York, New York 10003

United Kingdom Edition published by
ACADEMIC PRESS, INC. (LONDON) LTD.
24/28 Oval Road, London NW1

LIBRARY OF CONGRESS CATALOG CARD NUMBER: 75-29538

ISBN 0-12-134450-9

PRINTED IN THE UNITED STATES OF AMERICA

Contents

List of Contributors

Robert S. Brodkey, Department of Chemical Engineering, The Ohio State University, Columbus, Ohio 43210

S. N. B. Murthy, Project Squid Headquarters, Thermal Science and Propulsion Center, Purdue University, West Lafayette, Indiana 47906

Edward E. O'Brien, Department of Mechanics, State University of New York at Stony Brook, Stony Brook, New York 11790

G. K. Patterson, Department of Chemical Engineering, 101A. Chemical Engineering Bldg., University of Missouri–Rolla, Rolla, Missouri 65401

L. L. Simpson, Union Carbide Corporation, Chemicals and Plastics, P. O. Box 8361, South Charleston, W. Virginia 25303

H. L. Toor, Carnegie–Mellon University, Carnegie Institute of Technology, Schenley Park, Pittsburgh, Pennsylvania 15213

Preface

In June of 1974, a symposium on "The Application of Turbulence Theory to Mixing Operations" was held at the Pittsburgh, Pennsylvania national meeting of the American Institute of Chemical Engineers. This review book is an outgrowth of that symposium which the Editor arranged at the request of J. Gray. In addition, a sixth paper has been added to round out the subject. More details on this are given in the Introduction.

In this camera-ready book, we have used a uniform, simple notation that follows the Introduction. References have been standardized and appear at the end of each chapter. Cross-references to other contributions have been made. It is hoped that these efforts will give the reader a more usable text.

Finally, this seems to be the appropriate place to thank those people who have contributed to the overall production of this effort. To each author, I owe my thanks for their enthusiastic response to my original request to be contributors to the symposium, and to their prompt production of a revised manuscript that was a result of each author commenting upon the others' work and the editor trying to find a more or less middle of the road. I must also thank Miss Carolyn E. Patch for typing the entire copy-ready manuscript you are reading, for many minor corrections which the editor missed, and for help in many ways far beyond the call of duty.

Introduction

ROBERT S. BRODKEY

Department of Chemical Engineering
The Ohio State University
Columbus, Ohio 43210

Turbulent mixing of materials with or without the presence of chemical reaction is recognized as an important research field because of the many early problems that confronted chemical engineers working with blending, chemical reactors and combustion processes. The complexities of turbulent motion, mixing and kinetics precluded making much more than crude (but workable) empirical approaches. However, the continued need to solve such problems and the present-day importance of understanding and modeling in such diversified areas as the environment, propulsion systems, chemical lasers, oceanographics, the atmosphere and even medicine make research in this area of turbulence attractive to many workers in many fields. The goal of all of this research is to obtain an understanding of the physics of turbulent motions, of the effect of these motions on the transport of an associated scalar field, of the effect of chemical kinetics and to combine all these aspects in the form of a model of the process. Clearly, for many of the practical applications, temperature effects cannot be ignored, and in some cases, even effects such as drag reduction will have to be considered.

Testifying to the current interest, in 1974 two conferences on mixing were held. In May there was a Project Squid Workshop at Purdue University on "Turbulent Mixing of

Non-reactive and Reactive Flows" (proceedings to be published), and in June there was a symposium on "The Applications of Turbulence Theory to Mixing Operations" held at the Pittsburgh national meeting of the American Institute of Chemical Engineers. This review book is an outgrowth of the latter meeting, but the influence of the former is certainly felt. A number of the authors of this volume attended the first meeting and one is Director of Project Squid and acted as the host.

In this book an effort has been made to provide an up-to-date summary of the present status of research on turbulent motion, mixing and kinetics. Each author has been given the assignment to review that area in which his expertise lies. These then have been composited in such a manner as to extend from the basic theory to industrially important applications. The reviews do not deal directly with the vast background of material on turbulence itself, but rather assume a foreknowledge of this. Turbulence is discussed, but in the context of mixing and reaction in scalar fields. Thus in this brief introduction, it would seem appropriate to mention where one might find some of the background material that would help when one is confronted with the details of representing turbulent flows. I, of course, prefer Chapter 14 of Brodkey (1967), which provides an introduction to statistical turbulence and develops the subject area enough to include some aspects of dispersion, mixing and chemical reaction. There are more recent and more detailed efforts available. A good example is Tennekes and Lumley (1972). In this there are excellent discussions on the basic turbulence and the transport of heat and mass. These books can, most likely, provide the reader with source material in those areas where he might most need background.

An overview of what the authors cover should be helpful to the reader. O'Brien in Chapter I starts with a classification of turbulent reacting systems. This is quite important when one investigates limiting cases, and Toor in Chapter III summarizes the possibilities in this context. In brief, the application of the concepts and techniques of turbulence theory to reactant mixing is reviewed by O'Brien. The theory is stressed and the 2nd-order, one-step, irreversible reaction is emphasized. The asssumptions and limitations are given and a brief review of passive mixing is presented. This latter is covered in much more detail by Brodkey in Chapter II. The application of theory is first made by O'Brien to the case where the reactants are initially premixed as can exist in premixed flame problems. A postulate of statistical independence is presented which presumes that the decay of concentration fluctuations due to mixing and due to reaction are independent. Asymptotic behavior and approximate solutions are discussed for this approximation. O'Brien then considers initially segregated reactants and, in principle, the lack of statistical independence. However, as pointed out by Toor, if one makes essentially the same assumption, the calculated results compare well with data, at least to a stoichiometric ratio of 3. In these latter calculations, the mixing was known and only the course of a 2nd-order reaction was determined. O'Brien, on the other hand, outlines a strategy for solution where the mixing itself must also be determined from turbulence parameters. Very approximate methods to establish the mixing are given by Brodkey, and Patterson in Chapter V uses some of these to make parallel numerical calculations of both mixing and reaction. The case of very rapid reactions is equivalent to the mixing problem and O'Brien covers the passive contaminant analog and gives some shear flow results. Toor and Brodkey treat this limiting case in some detail. Finally, O'Brien concludes with some comments on current problems.

In the second chapter, Brodkey considers mixing without reaction in several different devices and tries to test possible simple methods of estimating turbulent mixing. The testing of estimations requires experimental information and thus this is emphasized in this chapter. The geometry dictates the turbulence, this in turn controls the mixing and the mixing influences chemical reaction. A mechanistic picture of turbulent mixing and possible descriptive parameters are presented. Experimental results on turbulence and mixing are given for pipe line, stirred tank, and multi-jet reactor systems. The theory first deals with the turbulence itself. A brief discussion is given of microscale, macroscale, and energy dissipation; all of which must be known to make estimates of mixing that are presented next. Here the basic equations given by O'Brien (Chapter I) and by Toor (Chapter III) are mentioned, as are approximate spectra results on which the simple mixing theories are based. The theory of Beek and Miller and that of Corrsin are presented in some detail. The means of using and modifying these results for mixing calculations are given. Brodkey next uses these approximate methods to compare experiments to theory for pipe flows, mixing vessels, and the multi-jet reactor. A summary of the results shows the wide range of agreement. The chapter concludes with a perspective and an extended detailed outline of references that should be of help to the reader for all aspects of the mixing problem.

Toor, in Chapter III, is concerned with predicting the time average rate of a chemical reaction in a turbulent field. Briefly, limiting solutions to the equations combined with experiment allow one to bypass the relatively intractable turbulence problem and relate reaction rates directly to mixing. The general problem is introduced and the very rapid reaction limit considered in some detail. The one-dimensional tubular-type reactor is analyzed in order to establish design equations. Brodkey, in Chapter II, has considered the mixing and turbulent fields in the same reactor configuration. Toor then treats the very slow reaction limit and inlet conditions. Slower speed reactions are not highly sensitive to the mixing, so a rigorous theory relating mixing to reaction does not appear to be needed nor easily confirmed, at least for simple 2nd-order reactions. An invariance hypothesis is introduced which is essentially the same as the postulate of statistical independence of O'Brien in Chapter I. The hypothesis allows one to calculate reaction rates from a knowledge of the reaction velocity constant, mean velocity, and mixing, at least up to a stoichiometric ratio of 3. At higher ratios, the theory would no doubt fail and one would have a lack of statistical independence as suggested by O'Brien. Some data are available for fast gas reactions, and these are discussed both by Toor and Brodkey. The mixing in liquid and gas systems is the same in terms of a convected distance. The time to mix a gas is shorter, but because of the faster velocity for the same dynamic conditions, the time is less. Design and other models available are reviewed by Toor, but much more detail on these aspects of the problem can be found in Chapter V by Patterson. Some results are presented for a two-dimensional reactor configuration.

In summary, Toor concludes that even without a tractable statistical theory of turbulent mixing, *ad hoc* methods and experiments have allowed us to gain considerable insight into the non-premixed 2nd-order chemical reaction problem.

Murthy, in Chapter IV, has provided a general review of reactive flows away from boundaries in which he has put parts of Chapters I and III of O'Brien and Toor into the perspective of current interest in modeling. The various approaches are reviewed, and a discussion of the extension to turbulent combustion is given. After a general introduction to the problems of chemically reactive flows, including comments on the lack of agreement on what should be measured and a plea for the selection of just a few definitive flow configurations, he reviews the experiments that have been made involving shear layer mixing. The conditions and measurements that are usually made are summarized before a number of the experimental findings on turbulent mixing of shear flows is given. The importance of the actual structure of the mixing layer is emphasized, although such information is still too new to have had any impact on modeling of reactive systems. The problems associated with modeling reactive flows are discussed in terms of the superimposed effects of convection, diffusion and reaction dynamics, and the nonlinearities that these introduce are considered. The model equations, time constants associated with these and the scales are all treated. The various assumptions invoked to enable a solution are given, i.e., an independence principle, Toor's approach, and intermittency in mixing layers. Compressibility effects are briefly considered before a more detailed treatment of statistical continuum theories (Edwards, Hill's extension, O'Brien and Chung) are treated. The section on turbulent combustion points out what must be considered, contrasts laminar and turbulent combustion, and considers the modeling of turbulent flames.

Patterson, in Chapter V, brings the foregoing material closer to practical application. For reactor systems that may exist in any commercial process, such concepts as backmixing, residence time distribution, premixed reactants, non-premixed reactants, macromixing, micromixing and the possible definitions of the intensity of segregation are all important and are explained. The traditional methods of designing and scaling-up mixers, using overall similarity concepts and rules of thumb are briefly presented as background for a review of the various reactor modeling methods when the feed is either premixed or non-premixed. The premixed reactant problem is clearly contrasted to the more difficult non-premixed case, and, in all cases, the various available models are discussed and briefly explained. The main emphasis is, however, on the development and application of a hydrodynamic mixing model which can be applied to reactions in geometries such as a stirred-tank mixer or a tubular reactor. The basic balance equations and computer means of solution are presented. Results are presented for application of the model to tank-mixers (both turbine and propeller designs)with or without reaction. Of particular importance is the application of the model to the tubular reactor data presented by Toor in Chapter III. An effort is made to put the work into context with other models such as the coalescence-dispersion model and the more complex models reviewed by Murthy in Chapter IV. Some final comments are given on what can be used for design and scale-up of industrial problems.

In the final chapter (VI), Simpson reviews turbulence and actual industrial mixing operations. Clearly, turbulence theory has had an impact on the design of reactant mixers and tubular reactors. Although completely general design methods are not yet in sight, the knowledge gained can help the designer to understand basic mixing phenomena. Mixing in pipelines by injection through a tee is treated, as well as the inherent mixing available from pipeline turbulence. Some comparisons are made based on the material previously presented by Brodkey in Chapter II. It is pointed out that all piping systems can cause mixing, but that care must be taken to control the input streams. Slugs of one fluid following the other need extremely long lengths to effect any kind of mixing. In this respect, several types of pipe line injectors and mixers are considered in terms of special processing requirements. It is pointed out that flow-splitting devices, designed for blending highly viscous fluids, are often very effective turbulent mixers. An extensive treatment of jet mixing is provided. The basic concepts are reviewed and jet mixing in tanks and solids in suspension are both treated in some detail. Finally, consideration is given to the tubular plug flow reactor where entrance, exit, and stratification effects are discussed.

REFERENCES

Brodkey, R.S. (1967) THE PHENOMENA OF FLUID MOTIONS, Addison-Wesley Publishing Co., Inc., Reading, Mass.

Tennekes, H. and Lumley, J.L. (1972) A FIRST COURSE IN TURBULENCE, The MIT Press, Cambridge, Mass.

NOMENCLATURE

A,B	mass or concentration fraction or concentration in constant density systems
A_{Rj}	heat transfer area in segment (j)
A	cross-sectional area (subscript p, pipe; n, nozzle; j, jet)
$\overline{A}_c$	concentration along jet centerline
a,b	fluctuation ($A - \overline{A}$, $B - \overline{B}$, respectively)
a', b'	rms value ($\sqrt{\overline{a^2}}$, $\sqrt{\overline{b^2}}$, respectively)
a_j	jet radius
C	constant
C_D	drag coefficient
C_p	heat capacity
$C(\vec{r})$	correlation for concentration fluctuations
d	$= \sqrt{I_s}$, diameter (subscript f, o.d. of finger pipe; i, impeller; j, jet at distance z; m, mixer; n, i.d. of nozzle; O, pipe; p, particle; p, max, maximum particle; T, tank)
D	molecular diffusivity
D_T	total diffusivity, molecular plus turbulent
E	scalar spectrum function
ΔE	activation energy
F	fractional conversion = 1 - X, pump head
F_0	pump head at shut off
f	correlation, fanning friction factor
g	gravity

H	height, height of liquid in tank
ΔH	heat of reaction
h	heat transfer coefficient, plane jet slot width
I	cd interaction parameter
I_E	Evangelista's mixing intensity
I_s	intensity of segregation $(\overline{a^2}/\overline{a_0^2} = a'^2/a_0'^2)$
J	$= nA - B$
j	$= J - \overline{J}$
K	scaling ratio
K_n	no. velocity heads loss in nozzle
K_p	no. velocity heads loss in piping
k	reaction velocity constant, wave number
k_0, $k_{0,s}$	lower wave number cutoff values for velocity and scalar fields
L	characteristic length
$L(\ )$	an operator
L_f	macroscale
L_m	half mixing distance
L_s	scale of segregation
M	mass
M_n	momentum flux parameter
m_r	$= (\varepsilon/L_s^2)^{1/3}/kA_0$, relative mixing intensity
N	impeller rotation rate
$N(0)$	zero crossings

N_D	first Damkohler Number $(k\overline{B}_0 L_m/U)$
N_{Re}	Reynolds Number (LU_x/ν)
$N_{Re,j}$	jet Reynolds Number
$N_{Re,p}$	particle Reynolds number
N_{Sc}	Schmidt Number (ν/D)
N_{yH}	no. of velocity heads
n	stoichiometric coefficient
n_f	vortex shedding frequency
P	products, power
p	pressure
Q	volumetric feed or flow rate
q	$= (u_x^2 + u_y^2 + u_z^2)^{1/2}$, instantaneous turbulent kinetic energy
R	rate of reaction, gas constant
R_A	mass production rate of species A
r	radial distance (subscript i, orifice; 0, pipe)
$\vec{r}$	vector distance between two positions
r_T	radial distance at tank wall
r_x	separation distance in the x-direction
s	displacement of jet by cross-flowing stream
T	temperature
T_a	activation temperature
t	time

NOMENCLATURE

$\overline{U}$	time-averaged velocity (subscript c, axial jet centerline; j, mean jet; n, mean nozzle ; p, mean pipe; r, max, maximum radial velocity from radial wall jet; s, mean cross flow)
U_z	axial velocity
$\overline{U}_z$	time-averaged axial velocity
U_s	superficial velocity
U_t	terminal velocity
U_m	mean velocity at minimum cross section
U_x	flow direction velocity
$\vec{U}$	velocity vector
$\vec{u}$	velocity fluctuation $(\vec{U} - \overline{\vec{U}})$
u'_x	flow direction rms velocity fluctuation
U^*	friction velocity
U^*_m	friction velocity at minimum transport conditions
V	volume
W_B	molecular mass of B
w	mass flow rate (subscript j, at given axial position in jet; js, in jet/slot length; n, through nozzle; ns, through nozzle/slot length)
X	$= \overline{A}/\overline{A}_0$
$\vec{x}$	position vector
Y	dimensionless concentration
y	distance normal to a surface, perpendicular to center of plane jet

y	dimensionless concentration fluctuation
Z	$= z/L_m$
z	axial distance, distance from wall of radial wall jet (subscript c, length of potential core; l, axial length along a jet in a cross flow; n, nozzle-to-wall distance of radial wall jet)

Greek

α	void fraction
β	stoichiometric ratio $(\overline{B}/n\overline{A})$ also $\overline{\overline{=}}$ jet stoichiometric ratio $= \overline{B}_{jet\ exit}/\overline{A}_{ambient}$
$\hat{\beta}$	$= B_0/nA_0$
γ_j	$= j'/\overline{J}$
γ_m	$= a'_m/\overline{A}_m$
γ	strain rate parameter, interdiffusion of reactants
δ_L	viscous sublayer thickness
δ	boundary layer thickness, radial wall jet dimension
ε	turbulent energy dissipation rate/mass
ξ	mixer L/d ratio
η	$= 1 - \sqrt{I_s} = 1 - d$, fractional completion of mixing, efficiency
θ	total included angle of jet
θ_d	diffusion relaxation time
θ_m	mixing half life
θ_r	reaction half life
λ	microscale

NOMENCLATURE

μ	molecular viscosity
μ_t	turbulent viscosity
ν	kinematic viscosity
π	= 3.1416...
ρ	density
ρ_p	particle density
ρ_s	cross-stream fluid density
$\Delta\rho_c$	critical density difference
σ	interfacial tension
τ	time scale for mixing and/or reaction, shear stress
ϕ	distribution function, spectrum, correlation parameter for mixing in tanks
ψ^2	dimensionless product of reactant fluctuations $(\overline{ab}/\overline{ab}_0)$
Ω	intermittency
ω	minimum transport settling velocity/U_t

Subscripts

1, 2	jet set 1, 2
0	inlet, initial, total
i	species, inlet, inside
I, II	scales
ave	average
j	jet
k	reaction
m	mixing
max	maximum
n	nozzle

o	outside
p	particle, pipe
s	scalar
w	wall
A	species A
B	species B

Superscripts

'	rms value of fluctuating component
$\overline{}$	time average
0	tracer or pure mixing

Theoretical Aspects of Turbulent Mixing of Reactants

Edward E. O'Brien

Chapter I

Theoretical Aspects of Turbulent Mixing of Reactants

EDWARD E. O'BRIEN

Department of Mechanics
State University of New York
Stony Brook, New York 11794

I. INTRODUCTION

I.A. *CLASSIFICATION OF TURBULENT REACTING SYSTEMS*

In the most general sense, a turbulent reacting system is one in which any part of the system is in turbulent motion. The turbulence may be generated by energy produced in the reaction: for example, it may be bouyancy generated due to strong exothermicity, or it may be induced by external input of energy as with mechanical mixers. Therefore, any reacting system in which one component is a fluid (gas or liquid) comes under the umbrella of the title of this symposium volume and this review paper. One can list, as follows, one possible classification of reactive systems in which turbulent mixing may play a significant role.

1. gases
2. miscible liquids
3. immiscible liquids
4. liquids dispersed in a gas
5. solids dispersed in a gas
6. solids dispersed in a liquid

It is a well-established result of hydrodynamic theory that the first two classes exhibit similar dynamical behavior (Batchelor, 1967), characterized by Reynolds number, for many fluids of not too complex structure; namely, those that obey the Navier-Stokes equations. The major distinction in practice arises from the relatively large Schmidt number that liquids usually display. The other classes listed above introduce new parameters, such as surface tension or moduli of elasticity which generally complicate the physics of their behavior when in turbulent motion.

Superimposed on such a phase classification scheme there must be a classification based on the type of reaction taking place between species. These are inherently molecular in nature and, at that level, complex descriptions have been provided (Bunford and Tipper, 1969). Since it is widely believed that turbulence exhibits a minimum length scale [roughly the Kolmogoroff length (Castleman, 1974)] for the size of eddies and that this length scale is very much greater than molecular mean free paths in ordinary circumstances, it is customary to assume that the chemical kinetic processes for miscible phases are not directly disturbed by the turbulence. That is, if two species exist in certain concentrations at a *point* in the turbulence, they will instantaneously react at the same rate as they would in a chemically homogeneous mixture at rest in the same concentrations. The *point* is understood to have the usual continuum meaning with a dimension larger than the mean free path, but smaller than the length over which concentrations vary significantly on a macroscopic scale. Despite its inherently molecular nature, the concept is analogous to the one by which molecular diffusion in laminar flows is modeled by Fick's law. Such an assumption is widely used, not only for the mixing of fluid phases, but also in the study of aerosol formation in the atmosphere, for example (Castleman, 1974). Some kinetic theorists dispute such a dichotomy between long-range and short-range molecular effects; however, this assumption permits one to bring reactive mixing directly into the main stream of turbulence research and to use, without modification, existing classification from reaction kinetics.

From a practical point of view, the most useful single parameter for describing the role of turbulence on chemical reactions is a ratio of time scales [(Toor, 1969, see also Chap. III), a time scale τ_k characteristic of the kinetic scheme to a time scale τ_m characteristic of turbulent mixing]. In a single-step, irreversible reaction, the inverse of the reaction rate constant in conjunction with characteristic concentrations can represent the chemical reaction time; whereas, the time for the decay of fluctuations of a scalar field in the turbulence might represent the mixing time adequately. Three regimes suggest themselves:

1. $\tau_k/\tau_m >> 1$ The slow reaction
2. $\tau_k/\tau_m \simeq O(1)$ Moderate reaction rate
3. $\tau_k/\tau_m << 1$ The very rapid reaction

In case one, it is abundantly clear, at least when approximate statistical homogeneity applies, that turbulence will induce chemical homogeneity before any significant reaction will occur and the fluctuations in concentration of any species at a *point* will generally be negligible compared to the mean concentration in determining the rate of reaction.

In the second case, complex coupling between the turbulence and the reaction is to be expected even under statistically homogeneous conditions.

In the third case, the behavior depends crucially on the nature of the reaction as we shall see later. In particular, for multispecies reactions in which the species are not uniformly distributed in space, the progress of the reaction will be diffusion limited since molecules must first diffuse to the same *point* before they can react. It is the rate of molecular diffusion, enhanced by the turbulence through stretching of isoconcentration lines, which must control the rate of progress of the reaction (O'Brien, 1967).

I.B. *GENERAL ASSUMPTIONS AND LIMITATIONS*

Although there is some literature devoted to reactive turbulent mixing of the kinds classified in section I.A. (Castleman, 1974; Toor, 1969; O'Brien, 1971; Corrsin, 1964A), it is in a primitive state when viewed from the perspective of turbulence theory. In fact, it is only in the last decade that real progress has been made in reaching some theoretical understanding of reactive turbulent flows of gases or miscible liquids with Newtonian behavior (refs. above and Gibson and Libby, 1972). The remainder of this review will be limited to this situation. Furthermore, little progress has been reported on the extremely complex question of how a chemical reaction will modify the turbulence which carries it. We content ourselves with mentioning a reference to a recent thesis (Dopazo, 1973) in which the problem has been carefully formulated and which also includes a thorough survey of the present status of the field.

A final simplification we adopt is to treat only simple reactions; by which we mean essentially one-step, second-order, irreversible reactions. These reactions have received almost all of the recent research attention in this subject, both theoretically and experimentally (refs. cited previously). These simple reactions display many of the characteristics of more complex kinetic schemes, such as diffusion control at very rapid reaction rates; at the same time, detailed turbulent mixing considerations can be incorporated into their study. We may directly extend the results discussed here to more complex kinetic schemes in which such a reaction is the controlling step in determining the rate.

I.C. *THE SECOND-ORDER, ONE-STEP, IRREVERSIBLE REACTION*

We will emphasize in this paper the turbulent mixing and chemical reaction between two reactants undergoing a one-step, exothermic, irreversible reaction of the form

$$A + nB \rightarrow mP + \text{Heat} \tag{1}$$

where n and m are stoichiometric coefficients, A and B represent reactant species and P is a product species. It will be understood that the heat produced is not sufficient to modify the turbulent field significantly. We further assume that Fick's law relates the molecular diffusion velocity of each species to its instantaneous concentration distribution in space in the usual way, and that the instantaneous mass production term at a *point*, which will arise in the balance equations for each species, is identical to that given by homogeneous kinetics, as was discussed in section I.A.

A typical balance equation for species A in the reactive system described by Eq. (1) can be written (Dopazo and O'Brien, 1973)

$$(\partial A/\partial t) + \vec{U}\cdot\vec{\nabla}A = D_A\nabla^2 A + R_A \tag{2}$$

where A is the mass fraction of species A, $\vec{U}$ is the turbulent velocity field taken as incompressible, D_A is the diffusion coefficient for species A, and R_A is the mass production rate of species A.

Each of the other species, in our case B and P, will have analogous mass conservation equations which have to be satisfied simultaneously with Eq. (2). When temperature variations are significant in either D_A or R_A, it will be necessary to solve the equation for thermal transport as well (Dopazo and O'Brien, 1973). The difficulties encountered in predicting the effects of turbulent mixing become clear when Eq. (2) is examined term by term. Turbulent convection, represented by the term $\vec{U}\cdot\vec{\nabla}A$, and molecular diffusion, given by the Fickian term $D_A\nabla^2 A$, together represent turbulent mixing in its simplest form. There exists an extensive literature on this subject including a thorough recent review to which we refer the reader (Brodkey, 1973 and Chapt. II). In the next section, a very brief summary of current mixing theories is presented.

A typical mass production term R_A appears in the form

$$R_A = -k(T)AB \tag{3}$$

where k(T) is the Arrhenius rate constant given by

$$k(T) = (nA\rho/w_B)e^{-T_a/T} \tag{4}$$

Here A is the preexponential factor, ρ is a mean density for the mixture, w_B is the molecular mass of B, and T_a the activation temperature of the mixture.

The added complexities that reactions contribute to mixing can be summarized from Eqs. (2) - (4) as follows:

1. The appearance of B and R_A causes a direct coupling between the mass balance equations for species A and B, which are likely to be statistically strongly correlated (O'Brien, 1971).
2. Since A and B are both non-negative quantities, inequalities are introduced between their means and higher order moments which vastly complicate the difficulties of finding approximations.

For example, one can show (O'Brien, 1968) that $\overline{A^3} > \overline{A^2}^2 \, \overline{A^{-1}}$ for any non-negative random variable A, where the overbar indicates an average value. Furthermore, non-negative random variables cannot be Gaussian.

3. For situations in which the temperature dependence of the reaction rate k(T) is important, it is clear that Eq. (3) will present great difficulty, if only because the equation of thermal transport then becomes coupled into the system.

I.D. *PASSIVE MIXING*

Before discussing efforts to come to grips with the turbulent reaction problem represented by Eqs. (1) - (3), it might be well to summarize briefly the state of the art of turbulent mixing without reaction (Brodkey, 1973; Chapt. II goes into this in more detail). In classical turbulence parlance (Corrsin, 1953), the word mixing is used to describe situations in which the scalar field is approximately homogeneously distributed throughout a turbulent field; which is to say, that the mean temperature or concentration at each point in the flow is about equal to the mean at every other point; the same is true of other moments of the scalar field. With such a usage, mixing is contrasted to turbulent diffusion which is characterized by significant turbulent transport from regions with large mean values of the scalar field to regions with lower mean values. For example, a heated turbulent jet of air exiting into cooler ambient air would be called turbulent diffusion rather than mixing, even though, both bulk turbulent transport and local mixing are taking place simultaneously. In this paper we do not adhere to such a dichotomy. The word mixing is meant to cover both statistically homogeneous turbulence and shear flow turbulence with no limitation to roughly uniform scalar fields.

Theoretical work on the two types of mixing has tended to fall into distinct categories: homogeneous mixing and shear flow mixing. The first case, in which both the turbulence and the scalar field are uniformly distributed in the large, has already contributed directly to reactor design (Corrsin, 1957, 1964B). We should like to draw attention to recent theoretical progress (Herring and Kraichnan, 1972) achieved mostly by researchers in turbulence dynamics who continue to generate approximation schemes and numerical procedures giving more adequate descriptions of the role of inertia and diffusion. We cannot mention these in any detail, but in section III, a convenient one is used to generate some decay results. Because of the existence of this body of research on turbulence and mixing, it is helpful to relate the mixing-with-reaction problems as closely as possible to nonreactive mixing so that use can be made of these carefully wrought approximation schemes. For mixing in nonhomogeneous flows, a similar attitude is adopted.

Recent activity in the generation of approximate closures for shear flow transport of scalar fields (Donaldson and Hilst, 1972; Spalding, 1971) produced, in part, by concern with atmospheric pollution problems gives promise of useful applications to pipe flow reactors and other characteristically nonhomogeneous devices. In one particular case, to be outlined in section IV, the advent of sophisticated measuring techniques for random scalar fields has allowed an immediate predicition of the reactant and product species statistics for fast reactions at the exit of a jet (Lin and O'Brien, 1974).

II. INITIALLY PREMIXED REACTANTS

The role of premixing of reactants is essentially to ensure that in every volume element in the flow, species A and B will exist in the same ratio. For the sake of definitude, we will consider the species to be in stoichiometric proportions. The generalization to nonstoichiometric ratios is important in practice, but is a trivial extension to the theory of premixed reactants. We also assume that the molecular diffusivities of all species are equal. Available evidence, experimental (Vassilatos and Toor, 1965) and theoretical (O'Brien and Lin, 1972), suggests that, in strong turbulence, variations in the numerical values for the diffusivities of the two reactants has negligible effect on low order moments, such as the mean, and that the same effective molecular diffusivity can be used for both species.

For convenience here, we also ignore the temperature dependence of the reaction rate, recognizing that this cannot be done when the mixture temperature at any point is in the neighborhood of the activation temperature. The incorporation of such thermal effects has recently been accomplished for low speed homogeneous turbulent mixing (Dopazo and O'Brien, 1973).

Under these constraints, and incorporating the stoichiometric coefficients into the mass fractions, the conservation equation for species A can be written

$$(\partial A/\partial t) + \vec{U}\cdot\vec{\nabla}A = D\nabla^2 A - kA^2 \qquad (5)$$

which also holds when A and B are the same species, as in reactions of the kind $A + A \rightarrow P$.

An even simpler situation arises when species B is everywhere in overabundant uniform supply. Then we find the mass conservation equation is well approximated by the linear form

$$(\partial A/\partial t) + \vec{U}\cdot\vec{\nabla}A = D\nabla^2 A - kA \qquad (6)$$

Both the first order reaction , (Eq. 6), and the second order reaction, (Eq. 5), have received considerable attention (Corrsin, 1964) in homogeneous systems from the point of view of predicting the decay of mean quantities, designated by an overbar, such as $\overline{A}$ or $\overline{A^2}$, etc., and the spectral behavior of the scalar intensity $a' = \overline{a^2}$ in the Kolmogoroff equilibrium range of the turbulence.

We note (O'Brien, 1960) that if $A_m(\vec{x},t)$ is a solution of Eq. (6) when $k \equiv 0$ (no reaction) then $A(\vec{x},t) = A_m(\vec{x},t)e^{-kt}$ is a solution of Eq. (6) when $k \neq 0$ and hence

$$\overline{A} = \overline{A}_m e^{-kt} \qquad (7)$$

$$a' = a'_m e^{-2kt}$$

etc.

The subscript m is meant to suggest simple mixing and, in fact, this expression of the moments of the concentration of species A in terms of the moments of a non-reacting scalar field which is exact for first order reactions has suggested an approximation for non-linear, single species reactions such as those described by Eq. (5). As will be indicated in Chapt. III, there is some experimental evidence for this assumption.

II.A. *POSTULATE OF STATISTICAL INDEPENDENCE*

It has been postulated (O'Brien, 1969) that the decay of concentration intensity at a point in the spectrum is given by the product of intensity decay due to mixing and intensity decay due to reaction where each acts as statistically independent phenomena. In a homogeneous system, this leads to a simple statement

$$a'(t)/a'(0) = [a'_m(t)/a'(0)][a'_k(t)/a'(0)] \tag{8}$$

The symbol a is used to denote a concentration fluctuation, $a = A - \overline{A}$, and $a'(0)$ is the initial r.m.s. fluctuation level of the concentration. $a'_m(t)$ is the solution of Eq. (5) when the reaction rate $k \equiv 0$ and $a'_k(t)$ is the solution of Eq. (5) when both $\vec{U}$ and D are ignored, the so-called stoichastically distributed reactant problem. The arguments leading to Eq. (8) are complicated and approximate. It would be useful to have experimental evidence to support it. Such evidence would come from an experimental study of simple mixing under the same turbulent conditions as those in which reactive mixing is carried out. Again, as pointed out in Chapt. III, there is some indirect evidence of this up to a stoichiometric ratio of 3 since it has been shown that mixing is independent of reaction at least up to this ratio.

In terms of the time scale characterization mentioned in I.A., we note that Eq. (8) predicts that the faster of the two phenomena, mixing and reaction, will dominate the decay of fluctuations in such single species reactions. There is no sign of the phenomenon of diffusion control which, in fact, was lost when we specified premixing. All potential reacting molecules are in contact initially and turbulent mixing cannot effectively change that situation. The basic reason for this is associated with the fact that the mass production term kA^2 of Eq. (5) is not directly sensitive to length scale distortion by the turbulent strain field since there are no spatial derivatives in such a term. This behavior is in marked contrast to that of molecular diffusion which exhibits a double spatial derivative and is, therefore, strongly spectrally selective. We will see in section III that initially segregated reactants have a quite different behavior at moderate to rapid reaction rates.

II.B. *ASYMPTOTIC BEHAVIOR AND APPROXIMATE SOLUTIONS*

From a theoretical point of view, Eq. (5) is tractable because it has well defined asymptotic behavior in certain limits (O'Brien, 1968). In the very rapid reaction limit it reduces to

$$\partial A/\partial t = - kA^2$$

which has exact stochastic solutions, and therefore, exact mean and rms behavior in terms of the initial statistics of the concentration field. From these exact results, and the non-negativeness of A, it is possible to devise closures which give good approximations to the exact mean and rms solutions. The nonlinear reaction term can then be well approximated at the level of the first few moments and there only remains the necessity to approximate the turbulent mixing by one of the techniques mentioned in I.D.

One technique we have found convenient for homogeneous turbulent flows is due to Lee (1966) who had modified and simplified the Direct Interaction results of Kraichnan so that detailed decay computations can be carried out over a reasonably long time. It has not yet been applied to the single species reactions described here, but we have adapted it for use with more complicated two species reactions and some results from that application will be discussed in section III.

Predictions based on such an integration of the equations of mixing and reactions should be on surer theoretical grounds than more widely used homogeneous mixing theories (Brodkey, 1973; Beek and Miller, 1959) discussed in Chapt. II, especially in regards to the decay of scalar intensity.

III. INITIALLY SEGREGATED REACTANTS

When the reactants are not premixed, it is necessary to consider simultaneous mass balance equations for both species A and B. Instead of Eq. (5), we have the following pair of equations to consider (O'Brien, 1971)

$$(\partial A/\partial t) + \vec{U}\cdot\vec{\nabla}A = D\nabla^2 A - kAB \quad (9a)$$

$$(\partial B/\partial t) + \vec{U}\cdot\vec{\nabla}B = D\nabla^2 B - kAB \quad (9b)$$

where for simplicity we have taken the reaction rate k as constant and the stoichiometric coefficients as unity. A more general non-isothermal treatment has been presented elsewhere (Dopazo and O'Brien, 1973).

III.A. *THE LACK OF STATISTICAL INDEPENDENCE*

Let us consider the consequences of an initial distribution of A and B which have the same mean but which are not distributed identically in space. Figure 1 represents

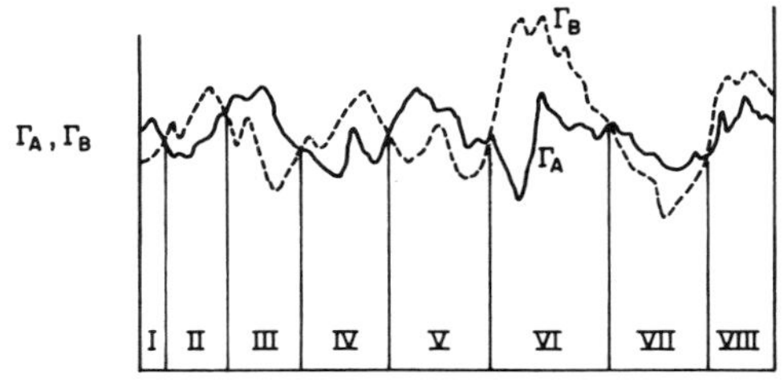

Fig. 1. Typical unpremixed distribution of species A and B with equal means ($\Gamma_A = A$, $\Gamma_B = B$).

the initial situation where the abscissa is any line in the flow and the ordinate the instantaneous concentrations of species A and B. Let us suppose that the reaction rate is so rapid that species A and B react instantaneously to form a product before the fluid can be distorted by the turbulence. In region I, for example, product would be formed and there would be remaining amounts of species A but no B. On the other hand, region II would consist of only species B with product and no A. Similarly for the other regions which we can represent as follows: Once the state pictured by Fig. 2

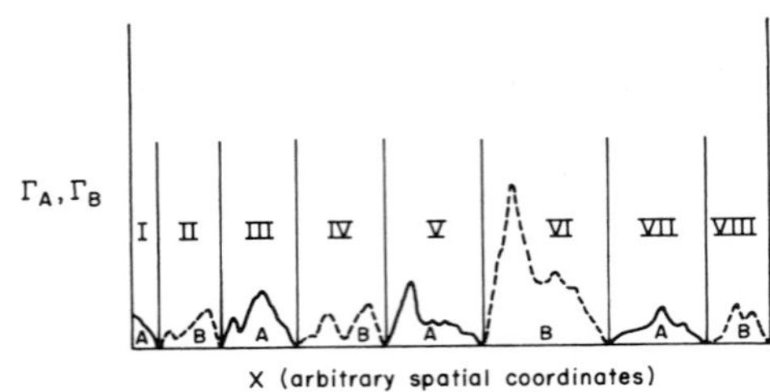

Fig. 2. Typical cross section for unpremixed very rapid reactions ($\Gamma_A = A$, $\Gamma_B = B$).

has been reached, further reaction can only take place between species A and B when molecular diffusion of each species brings it into contact with the other. Turbulent stirring of the reacting mixture increases the surface area of contact between regions of species A and regions of species B and in that way enhances the rate at which reaction occurs. The general state represented by Fig. 1 is what we mean by unpremixed reactants. If a similar picture were drawn for the premixed reactants of section II, it would show A and B randomly distributed in space but at every point A equals B (for stoichiometric proportions) and the state pictured in Fig. 2 is never reached. Figure 2 is a representation of what is termed *segregated* reactants. Such segregation more often arises from the method of injection of reactants.

In general, the reaction rate is not so fast as to cause the instantaneous development of segregated reactants from unpremixed reactants and there is, instead, simultaneous reaction and distortion of fluid volumes by the turbulence. It is clear that statistical independence cannot be possible for initially segregated reactants. Suppose the molecular diffusivity were identically zero then no matter how strong the turbulence or how effectively it stirs the scalar fields, there can never be further reaction beyond the state represented by Fig. 2 since the species never come together. Chemical reaction is then entirely eliminated even though both species are present in the fluid. Hence, the rate at which the reaction continues is not independent of the nature of the turbulence, and an equation like Eq. (8) is clearly incorrect. We will come back to the very rapid reaction in section IV. Let us now ask how a reaction described by Eqs. (9a) and (9b) might be approximated, realizing that no simple result analogous to Eq. (8) is possible for initially segregated reactants unless the reaction rate is so slow that mixing can remove the segregation of species before any significant reaction occurs.

III.B. *A STRATEGY FOR SOLUTIONS*

The details of this procedure are quite cumbersome and, as they are available elsewhere (Lin and O'Brien, 1972), we will not repeat them here. The strategy is as follows: The equation describing the behavior of either species represents three distinct physical effects: turbulent convection, molecular diffusion and chemical reaction. Each of these three phenomena, taken pairwise, are modeled in a way which preserves all that we know about them. For example, turbulent convection and molecular diffusion have long been studied as turbulent mixing and there exist techniques to model that behavior as we mentioned in section I.D. We have chosen to use a modification (Lee, 1966) of the Direct Interaction hypothesis (Orszag, 1970). It appears to give an adequate representation of mixing and is sufficiently simple compared to Kraichnan's original work (Kraichnan, 1959) that it has been possible to continue the computations of spectral decay over a significant fraction of the mixing life time. The technique is well suited to numerical computation and no unphysical results have been obtained, in contrast to the unfortunate behavior (O'Brien, 1960) of zero cumulant approximations. For a more detailed review, the reader should consult the original paper and its application to unpremixed reactants. Other more ambitious approximations could be used but the penalty is expensive computer requirements or a shorter time history.

The second pair of phenomena involves turbulence and chemical reaction without diffusion. The result that the reaction should proceed for such a system at a rate independent of the turbulence (Hill, 1970) leads to strong constraints on the nature of approximations to the chemical reaction terms. Specifically, the reaction should behave as it would were there no velocity field. We return to a situation, similar to the one mentioned in section II.A., where the equations to be solved are greatly simplified.

$$\partial A/\partial t = - kAB \qquad (10a)$$

$$\partial B/\partial t = -kAB \tag{10b}$$

Exact solutions of this pair can be obtained and used to test approximations for the mass production term (Lin and O'Brien, 1972; Hilst, 1973).

The final pair, molecular diffusion with chemical reaction in the absence of turbulence, cannot be solved exactly (Lin, 1971), but a great deal is known about its behavior. Furthermore, numerical integration of Eqs. (9a) and (9b) with the turbulence term removed is a simple matter and, again, a good testing ground for concentration field approximation is provided.

Finally we put together the mixing approximation and the reactive field approximation to satisfy all of the conditions we know of for the three phenomena taken pairwise and we hope that the combined approximation will adequately describe the three phenomena taken simultaneously. The ultimate justification of this method is in a comparison of the predicted results with experimentally observed behavior. We know of no existing experimental data with which to compare our computations at this time. Our expectation for its success rests on the lack of a spatial operator in the reaction term so that the effectiveness of turbulence in distorting the shape of blobs of concentration should not directly alter the instantaneous reaction rate. The detailed equations which result from these approximations are derived and numerically integrated in Lin (1974). Alternate methods for obtaining approximate solutions to the equations of this section have been proposed by several authors (Donaldson and Hilst, 1972; Hilst, 1973: Chung, 1973).

IV. VERY RAPID REACTIONS

The role of very rapid reactions was suggested by the discussion in section III. In fact, if the reaction is everywhere very rapid, as defined in section I, one can expect a picture such as Fig. (2) to represent the state of affairs in the reacting mixture for all times. Diffusion of one species from its own territory into that of its neighbor leads to immediate annihilation and the formation of very thin reaction zones separating regions of segregated species (Gibson and Libby, 1972; Hawthorne *et al.*, 1949). The analogous laminar flow situations have been reviewed recently by Williams (1971). Professor Toor (1965, 1969), who has contributed fundamentally to our understanding of this situation in turbulence, has reviewed this theory in Chapt. III of this volume. We will restrict ourselves to describing a method of attack for very rapid reactions which appears to have great promise even in shear flows. In section IV.B., we present some computed results for a very rapid reaction in the shear layer produced by a jet.

IV.A. *THE PASSIVE CONTAMINANT ANALOG*

Equations (9a) and (9b) can be written in an alternate form if we let $J = A - B$ and define the operator L by

$$L(\ ) = [(\partial/\partial t) + \vec{U}\cdot\vec{\nabla} - D\nabla^2]$$

(the mixing operator)

then

$$L(A) = -kA(A - J) \quad (11a)$$

and

$$L(J) = 0 \quad (11b)$$

Very rapid reactions with unpremixed reactants are known to proceed at a time scale of the order of the time scale of turbulent mixing. It follows that $kA(A - J)$ is finite almost everywhere as $k \to \infty$. We note that such a term cannot be zero since taking the mean of Eq. (11a) in a statistically homogeneous flow we find

$$\partial \overline{A}/\partial t = -\overline{kA(A - J)}$$

and $\overline{A}$ must decay at finite rate until all the reactants are expended. Therefore

$$\overline{A(A - J)} = 0\ (1/k) \text{ as } k \to \infty$$

and for very large k we obtain the result

$$\begin{aligned} \text{Either } A &= 0 \qquad B = -J \\ \text{or } A &= J \qquad B = 0 \end{aligned} \tag{12}$$

Relationship (12) allows us to construct the properties of both A and B if J is known (O'Brien, 1971). But from Eq. (11b), J is described by a mixing equation such as would be obeyed by any nonreacting passive contaminant, for example, temperature. This suggests that from an experiment on the mixing of temperature, one should be able to deduce information about a rapid chemical reaction in the same turbulence. This turns out to be so, provided the information on temperature is on its probability density function in the turbulence. Conversely, one can establish a description of mixing by measuring conversion of rapid reactions as done by Toor and further described in Chapt. III. The equivalence is shown there and in Chapt. II where the corresponding mixing measurements are described.

IV.B. *SOME SHEAR FLOW RESULTS*

Measurement of the probability density function of temperature at a cross section of a heated jet has been made by Tutu and Chevray (1973) and converted, by the analogy mentioned in IV.A., into reactive species information (Lin and O'Brien, 1974). They have obtained measured probability density profiles of temperature at 8 radial locations in a plane 15 diameters downstream of the exit and at several locations along the axis of symmetry of the jet. From their measurements, the statistical quantities related to A, B and the product P have been computed where species B is emitted by the jet and species A is in the ambient fluid. For example, one can predict $\overline{A}$, a', $\overline{B}$, b' and $\overline{P}$ as functions of position with a jet stoichiometry ratio, β, as a parameter. By definition, $\beta = \overline{B}$ at the jet exit/ $\overline{A}$ ambient. The details are contained in Lin and O'Brien (1974) along with detailed plots of the low order concentration moments at a cross section 15 diameters downstream from the jet exit. Some centerline concentration predictions at 10, 15, 20 and 25 diameters downstream as obtained from measurements in a heated jet with the same geometry and Reynolds number have also been computed but not previously published. Figure 3 shows how centerline values of $\overline{A}$, $\overline{B}$, $\overline{P}$, a' and b' vary with axial position when $\beta = 1$. Figure 4 presents similar results for the mean concentration when $\beta = 10$. The jet Reynolds number based on exit diameter was 31.6×10^4, the exit temperature $55^{\circ}C$, the environmental temperature $29.9^{\circ}C$ and the jet diameter 9 in. Similar measurements of probability density functions have been made on boundary layers (Zaric, 1972) and in the wakes of cylinders (LaRue and Libby, 1974). Temperature and injected helium (Stanford and Libby, 1974) have both been used as passive scalar nonreacting contaminants and there seems to be no reason why similar work on model reactors could not be carried out rather simply. Singh (1973) has reported measurements of centerline mean

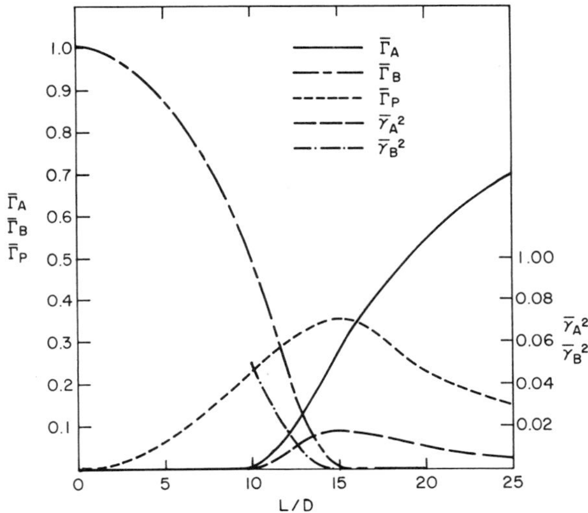

Fig. 3. Jet centerline concentration profiles versus dimensionless downstream distance, $\beta = 1$ $(\overline{\Gamma}_A = \overline{A},\ \overline{\Gamma}_B = \overline{B},\ \overline{\gamma_A^2} = a',\ \overline{\gamma_B^2} = b'$ *and* $\overline{\Gamma}_p = \overline{P})$.

concentrations in the near field of a reactive jet. The Reynolds number is much smaller than that used by Tutu and Chevray (1973), so direct comparison with the predictions of Lin and O'Brien (1974) are not possible. However, Singh observes the same qualitative behavior as that predicted by the analogy. He also explored the possibility of predicting moderate rate reactions from very rapid reaction data and, in a sense, used the analogy backwards to predict nonreactive mixing from rapid reaction data. These aspects will be discussed further in the next two Chapters.

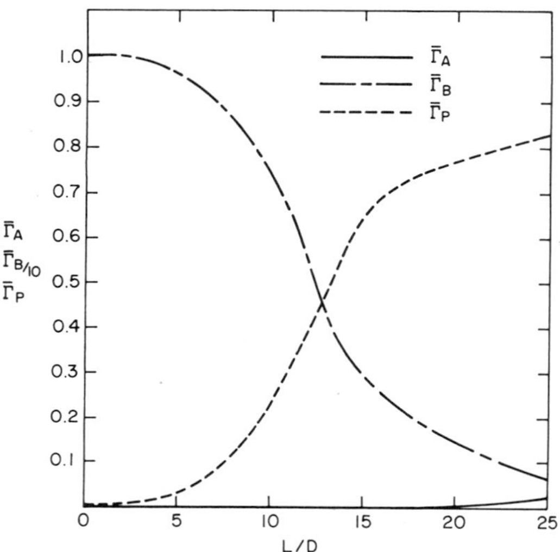

Fig. 4. Jet centerline mean concentration profiles versus dimensionless downstream distance, $\beta = 10$, $(\overline{\Gamma}_A = \overline{A}$, $\overline{\Gamma}_B = \overline{B}$, $\overline{\gamma_A^2} = a'$, $\overline{\gamma_B^2} = b'$ *and* $\overline{\Gamma}_p = \overline{P})$.

V. CURRENT PROBLEMS

The appearance of the probability density function in reactive mixing of very rapid reactions has led to a reformulation of reactive flows with an emphasis on these functions rather than moments (Dopazo, 1973). This formulation seems to be more natural, at least in flows with strong thermal effects and fast chemistry. Current theoretical work is aimed at formulating useful approximations (Lundgren, 1972; Fox, 1973; Dopazo and O'Brien, 1975) in this language, and one can hope to see some progress in this direction in the near future. The analogy described in section IV promises a fruitful partnership between relatively simple measurements in nonreacting turbulent mixers and theoretical work which can convert those measurements into detailed information about fast reactive fields in the same turbulence. There is every hope that a similar technique can be applied, in part, to moderate rate reactions as well (Singh, 1973). The possibility of the thermal field produced by the reaction modifying the mixing characteristics significantly is a difficult aspect of the general reactive mixing problem which is sure to receive more attention in the near future. It seems likely that simple moment closures will not be adequate in such cases (O'Brien, 1974; Donaldson, 1974). There is a lack of detailed experimental data on chemically reacting flows due undoubtedly to the great difficulties encountered in designing probes of sufficient stability and with sufficient spatial and temporal resolution. Such data has a vital role to play in testing approximations and in forming intuitive pictures as we attempt to apply turbulence theory to the mixing of reactant species.

VI. ACKNOWLEDGMENTS

The author is grateful to the National Science Foundation which, through Grant K040738, has supported his research in this area.

VII. REFERENCES

Batchelor, G.K. (1967) AN INTRODUCTION TO FLUID DYNAMICS, Camb. Univ. Press, London.

Beek, J.,Jr. and Miller, R.S. (1959) *Chem. Eng. Progr. Symposium, Ser. No. 25, 55,* 23.

Brodkey, R.S. (1973) in FLUID MECHANICS OF MIXING, 1, ASME.

Bumford, C.H. and Tipper, C.F., Editors (1969) COMPREHENSIVE CHEMICAL KINETICS, Vol. 1-13, Elsevier Press, Amsterdam.

Castleman, A.W. (1974) in SPACE SCIENCE REVIEWS, in press.

Chung, P.M. (1973) *Phys. Fluids 16,* 1646.

Corrsin, S. (1953) in PROCEEDINGS OF FIRST IOWA THERMODYNAMICS SYMPOSIUM, Iowa State University, Ames.

Corrsin, S. (1957) *A.I.CH.E.J. 3,* 329.

Corrsin, S. (1964A) *Phys. Fluids 7,* 1156.

Corrsin, S. (1964B) *A.I.CH.E.J. 10,* 870.

Donaldson, C. du P. (1974) presented at the Workshop on Turbulent Mixing, Project Squid, Purdue Univ., West Lafayette, May 20-21.

Donaldson, C. du P. and Hilst, G.R. (1972) *Environmental Sci. and Tech. 6,* 812.

Dopazo, C. (1973) Ph.D. Thesis, State University of New York at Stony Brook.

Dopazo, C. and O'Brien, E.E. (1973) *Phys. Fluids 16,* 2075.

Dopazo, C. and O'Brien, E.E. (1975) *Phys. Fluids* (to appear).

Fox, R.L. (1973) *Phys. Fluids 16,* 957.

Gibson, C.H. and Libby, P.A. (1972) *Comb. Sci. and Tech. 6,*29.

Hawthorne, W.R., Weddell, D.S. and Hottel, H.C. (1949) 3RD SYMPOSIUM ON COMBUSTION FLAME AND EXPLOSION, 266. The Williams and Wilkens Company, Baltimore.

Herring, J.R. and Kraichnan, R.H. (1972) LECTURE NOTES IN PHYSICS, *12,* Berlin, Springer.

Hill, J.C. (1970) *Phys. Fluids 13,* 1394.

Hilst, G.R. (1973) AIAA Paper 73-101, New York.

Kraichnan, R.H. (1959) *J. Fluid Mech. 5,* 497.

LaRue, J. and Libby, P.A. (1974) *Phys. Fluids* (to appear).

Lee, J. (1966) *Phys. Fluids 9,* 1753.

Lin, C-H. (1971) M.S. Thesis, State University of New York, Stony Brook.

Lin, C-H. (1974) Ph.D. Thesis, State University of New York, Stony Brook.

Lin, C-H. and O'Brien, E.E. (1972) *Astronautica Acta 17,* 771.

Lin, C-H. and O'Brien, E.E. (1974) *J. Fluid Mech. 64,* 195.

Lundgren, T.S. (1972) LECTURE NOTES ON PHYSICS 12, Springer, Berlin.

O'Brien, E.E. (1960) Ph.D. Thesis, The Johns Hopkins Univ.

O'Brien, E.E. (1968) *Phys. Fluids 11,* 1883.

O'Brien, E.E. (1969) *Phys. Fluids 12,* 1999.

O'Brien, E.E. (1971) *Phys. Fluids 14,* 1326.

O'Brien, E.E. and Lin, C-H. (1972) *Phys. Fluids 15,* 931.

O'Brien, E.E. (1974) presented at the Workshop on Turbulent Mixing, Project Squid, Purdue Univ., West Lafayette, May 20.

Orzag, S. (1970) *J. Fluid Mech. 41,* 363.

Singh, Madan, (1973) Ph.D. Thesis, Carnegie-Mellon Univ., Pittsburgh.

Spalding, D.B. (1971) 13th Symp. on Combustion, Combustion Institute, Pittsburgh, 649.

Stanford, R. and Libby, P.A. (1974) *Phys. Fluids* (to appear).

Toor, H.L. (1969A) *Ind. Eng. Chem. Fund. 8,* 655.

Tutu, N.K. and Chevray, R. (1973) *Bull. Am. Phys. Soc. 18,* 11.

Vassilatos, G. and Toor, H.L. (1965) *A.I.Ch.E.J. 11,* 666.

Williams, F.A. (1971) ANNUAL REVIEW OF FLUID MECHANICS, 3, Annual Reviews, Inc., Palo Alto.

Zaric, A. (1972) C.R. *Acad. Sci. Paris, t. 275, Serie A,* 459.

Mixing in Turbulent Fields

Robert S. Brodkey

Chapter II

Mixing In Turbulent Fields

ROBERT S. BRODKEY

Department of Chemical Engineering
The Ohio State University
Columbus, Ohio 43210

I. ABSTRACT

The principles of mixing on a macroscopic level have been known and used for a number of years for the scale-up of mixing processes. However, these scale-up methods have been empirical in nature due to the lack of information about mixing theory on a microscopic level. The complexity of problems involving turbulence, such as mixing, has made a quantitative description difficult. However, the statistical turbulence approach has led to a much deeper understanding in the field of mixing and related turbulent scalar transport processes. The analysis allows us to define measurable mixing criteria. A knowledge of the parameters of the turbulence provides the information necessary to estimate the degree of mixing. The parameters can often be estimated from the geometry of the flow system and simple empirical relationships.

II. INTRODUCTION

The term mixing is a loose one, and encompasses nearly as many definitions as there are workers in the field. This work will deal with localized or fine scale mixing, i.e., mixing that we would like to have when we are interested in encouraging chemical reaction between two species. We would like to know how the molecular diffusion of the individual species and the violent turbulent motions interact to bring the molecules together for reaction to occur. Turbulence is a difficult subject; it is still more difficult to combine the effects of turbulence and molecular diffusion and, to incorporate kinetics is a further complication which makes the analysis most difficult. Thus, it is not surprising that approaches to this subject involve idealized models.

As a brief summary of what is to come, we will attempt to establish the criteria by which we can measure the degree of mixing. The experimental systems used and the type of data obtained in making such measurements will provide some insight into the problems encountered. A review of theoretical efforts on the turbulent mixing problem will indicate the type of predictions that can be made. Of course, a question always of interest is the comparison of the experimental results with theory. Kinetics will not be incorporated into the problem as this will be reserved for other papers, although mention will have to be made of the very rapid reaction limit since this can be directly compared to the mixing. All of this is necessary background to our understanding of the turbulent mixing mechanism. What we would like to be able to do is pictured in Fig. 1. From a known geometry we would like to be able to predict or, if necessary, measure the parameters of turbulence. From this we want to obtain the mixing. Then, with incorporation of kinetics, we want to predict the full range of mixing with chemical reaction from the slow self mixing (back mixing) to the fast reaction limit.

II.A. *THE MIXING PROCESS*

We are interested in what we have just called fine scale mixing. It will help to put this type of mixing into the context of mixing in general. Our idea of mixing depends, to a great extent, on our definition of the term *mixture*. We shall use mixing to mean any blending into one mass, and

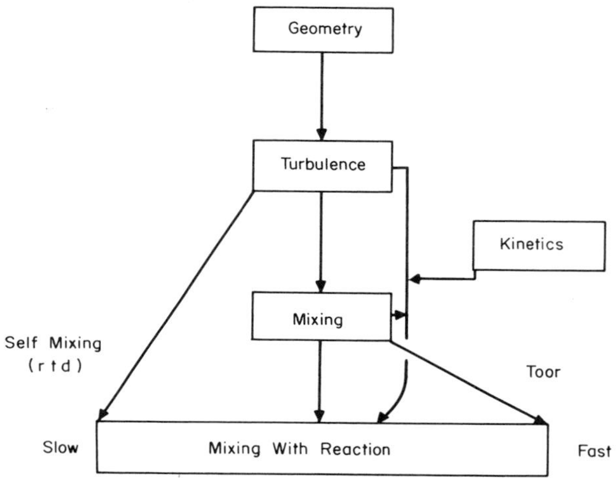

Fig. 1. Turbulent motion, mixing and kinetics

mixture to mean a complex of two or more ingredients which do not bear a fixed proportion to one another and which, however thoroughly commingled, are conceived as retaining a separate existence. In any specific case, our mixture will depend on the scale of our view. If our scale of view is very large, then even a very coarse mixture may be all we want, and molecular diffusion means little to us. Here, mixing, blending and dispersion are all equivalent. But it we want a mixture on a finer scale, then dispersion will not do alone, and we must have fine scale mixing by the smallest of eddies in the turbulence in conjunction with molecular processes. We will have to look at each item in even more detail if we are to obtain a real feel for the mixing process.

There is a paradoxical point that should be mentioned first. The mixing of two individual species (including molecular diffusion effects as we must) can give rise to a mixture of the species since the molecules of the two species can always be distinquished. The same is not true for heat or for momentum since the final blend of these are completely homogeneous, in that, at the end of mixing, we cannot distinquish a hot element of fluid or an element with a high velocity from one that had been cold or had a different velocity. We will not let this bother us since the basic processes are quite similar, especially for the scalar quantities of heat and mass, and all are often referred to as mixing processes.

Let us proceed by briefly discussing each of the processes that are usually considered as one form or another of diffusion and mixing. The word *diffusion* means the act of spreading out; it connotes nothing of the mechanism providing the spread. However, the unmodified work usually implies diffusion by molecular means. Since other processes can give rise to diffusion and, indeed, are called *diffusion processes*, the term *molecular diffusion* is better used to signify diffusion caused by relative molecular motion. In turbulent flow, there is bulk motion of large groups of molecules. These groups are called *eddies* and give rise to the material transport called *eddy diffusion* or dispersion. Nonmolecular and noneddy diffusion processes can be grouped into a class which will be called *bulk diffusion*. In each case, there is some bulk motion giving rise to diffusion which is superimposed on either molecular or eddy diffusion or both. For example, in turbulence, the problem is complicated by three superimposed diffusional processes. Molecular diffusion is always occurring and may not be always neglected. Superimposed is the gross random eddy motion, causing the eddy diffusion. Finally, it is possible to have other types of bulk diffusion occurring simultaneously, such as axial diffusion. It is this latter combination of molecular, eddy and bulk effects that is best described by the term *mixing*.

Molecular diffusion is a product of relative molecular motion. In any system where there are two kinds of molecules, if we wait long enough, the molecules will intermingle and form a uniform mixture on a submicroscopic scale (by submicroscopic, we mean larger than molecular, but less than visual by the best microscope). This view is consistent with the definition of a mixture for we know that if we were to use a molecular scale, we would still observe individual molecules of the two kinds, and these would always retain their separate identities. The ultimate in any mixing process would be this submicroscopic homogeneity where molecules are uniformly distributed over the field; however, the molecular diffusion process alone is generally not fast enough for present-day processing needs. In some systems, molecular diffusion is so slow as to be completely negligible in any reasonable finite time; high molecular weight polymer processing is a good example of this state.

If turbulence can be generated, then eddy-diffusion effects can be used to aid the mixing process. For some materials, the generation of turbulence would be too expensive because of high viscosity, and, in others, it might be impossible because of product deterioration under the high energy inputs required.

Each of the bulk-diffusion phenomena tends to reduce the scale by spreading a contaminant over a wider area. The molecular diffusion is enhanced because of the larger area. It is important to note that if the molecular diffusion is rapid enough, the system may be almost submicroscopically mixed by the time the bulk diffusion has spread the contaminant over the field. With low rates of molecular diffusion, this will not be true.

II.B. *THE MECHANISM OF MIXING*

The type of mixing of concern is the mixing between two or more streams. We have component A in one stream and component B in the other. We want to mix these into a homogeneous system where molecules of A and molecules of B can come into contact. The smallest scale of turbulence is large compared to molecular size, and the smallest eddy contains millions of molecules. Turbulence can only reduce the size to that of its smallest eddy, and thus we would still have many A molecules in one region and many B molecules in an adjacent region. One must have molecular diffusion in order to bring the A and B molecules together to react. This is the interaction between turbulent motion and molecular diffusion. This mixing should be contrasted to another type often referred to as a back-mixing, or better as self-mixing. Self-mixing is the mixing of a homogeneous fluid entering a mixing vessel. Here, material recently introduced is mixed with material that has been in the vessel for a longer period of time. This is a mixing in time and not a mixing between components. As it is usually formulated, this mixing does not involve molecular diffusion.

Next, let us consider a model of the details of the process as the system proceeds from an unmixed state to a mixed one. Let us take two streams that are in some way separately identifiable. Figure 2 illustrates the initial

Fig. 2. Dispersion

process of *dispersion,* which is defined as the act of breaking apart and causing to go different ways. Turbulent dispersion does *not* necessarily involve molecular diffusion processes at all; in fact, the molecular diffusivity does not even enter the dispersion problem as described by Taylor (1921); its effect was added by others later. A uniform dispersion or mixture can be obtained and one can continue to reduce the size of the blobs until they can be reduced no further by the action of the turbulence which does have a size below which it is ineffective. At the same time, molecular diffusion is usually acting as shown in Fig. 3. The final, very fine scale,

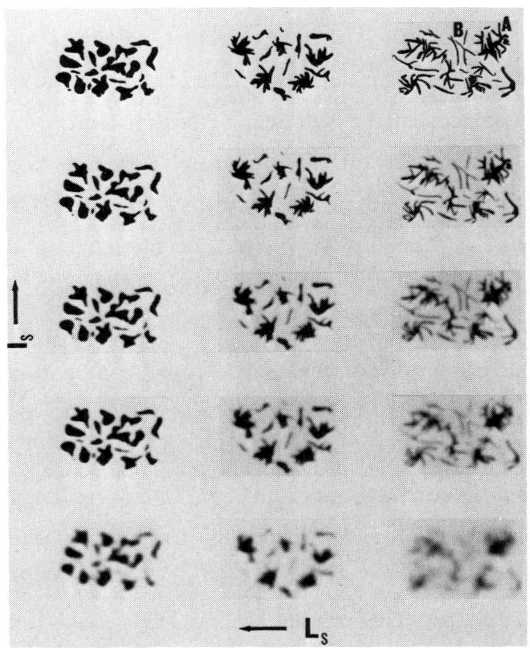

Fig. 3. Mechanism of mixing

uniform mixture must come from the molecular mixing provided by diffusion.

There are two criteria which, for now, we can treat qualitatively and define more quantitatively later. The first is the *scale of segregation* and is a measure of the size of unmixed clumps of pure components. This is a measure of some average size. As the clumps are pulled and contorted, the scale of mixing is reduced; this would be going from left to right along the top of Fig. 3. The second criteria is the *intensity of segregation*, which describes the effect of molecular diffusion on the mixing process. It is a measure of the difference in concentration between neighboring clumps of fluid. The intensity, for each value of the scale, is illustrated by the columns in Fig. 3.

To summarize, the turbulent process can be used to break up fluid elements to some limiting point; however, because of the macroscopic nature of turbulence, one would not expect the ultimate level of breakup (scale) to be anywhere near molecular size. Since energy is required for this reduction in scale, the limiting scale should be associated with the smallest of the energy containing eddies. This size will be large when compared with molecular dimensions. No matter how far we reduce the scale, we still have pure components. Depending on the size observed, any one of these levels in scale might be considered mixed; however, from a view of submicroscopic homogeneity, where molecules are uniformly distributed over the field, none is mixed. Without molecular diffusion, this ultimate mixing cannot be obtained.

Molecular diffusion allows the movement of the different molecules across the boundaries of the elements, thus reducing the difference between elements. This reduction in intensity will occur with or without turbulence; however, turbulence can help speed the process by breaking the fluid into many small clumps, and thus allowing more area for molecular diffusion. When diffusion has reduced the intensity of segregation to zero, the system is mixed. The molecules are distributed uniformly over the field. Various degrees of this combined process are shown in Fig. 3. In systems where reaction is to occur, the need for submicroscopic mixing is apparent, for without it, the only chemical reaction that could occur would be on the surface of the fluid clumps. Danckwerts (1958) has discussed the importance of this degree of mixing of two reactants; the intensity of segregation must be reduced rapidly so as to avoid local spots of concentrated reactant and the usually associated undesirable side reactions. In jet mixing, the scale of segregation is reduced by eddy motion while the molecular diffusion reduces the intensity. In a jet, if a solid product occurs, its particle size will be a function of the rate of reduction of segregation. The same would be true in the quenching of a jet of a hot gas reaction mixture; the freezing of the reaction products will depend on the reduction of segregation. Another example is the jet flame, where the oxygen is obtained from the surrounding air. The flame will depend on the segregation of the two gases. In laminar flames, the mixing is poor because the scale of segregation is high. The flame occurs along a surface and is controlled by the molecular diffusion across that surface. In a turbulent flame, eddy diffusion will reduce the scale and provide more area for molecular diffusion and thus more contacts for burning.

II.C. *CRITERIA FOR MIXING*

Two aspects of mixing have been mentioned: the degree to which the material has been spread out by the turbulent action (scale of segregation) and the approach to uniformity by the action of molecular diffusion (intensity of segregation). Danckwerts (1953) has provided definitions for these in terms of the statistical variables that can be measured in the mixing field. The criteria cannot be applied where gross segregation occurs. This restriction implies that the concentrations are grossly uniform over the field, but not necessarily uniformly mixed on anything but the largest scales.

II.C.1. Scale of Segregation

The main parameter of mixing is the instantaneous concentration fraction (A). Of importance is its average ($\overline{A}$) and the root mean squared fluctuation (a') about its average:

$$A = \overline{A} + a$$
$$a' = \sqrt{\overline{a^2}} \tag{1}$$

The criteria of mixing can be defined in terms of these. An Eulerian statistical concentration correlation function which is a measure of the similarity of concentration fluctuations at two points in space can be defined as

$$C(\vec{r}) = \overline{a(\vec{x})a(\vec{x} + \vec{r})}/a'^2 \tag{2}$$

The scale of segregation is defined in terms of the correlation, and is an integral relation:

$$L_s = \int C(\vec{r})d\vec{r} \tag{3}$$

The scale of segregation is an average over relatively large values of $\vec{r}$, and thus is a measure of the large-scale breakup process, but not of the small-scale diffusional process. In liquids with very slow molecular diffusion, the scale would decrease to some limiting value dependent upon the distribution of globs as caused by the turbulent field. This could happen before the fine scale mixing by molecular diffusion progresses very far. In gases, where molecular diffusion is rapid, the scale may not be reduced appreciably before diffusional effects become important.

II.C.2. Intensity of Segregation

The intensity of segregation is defined as

$$I_s = \overline{a'^2} / \overline{A}\,\overline{B} = \overline{a'^2}/\overline{a_0'^2} = d^2 \qquad (4)$$

and is measured at a *point* for a long enough time to obtain a true average. The subscript 0 refers to the initial value. The intensity of segregation is unity for complete segregation (i.e. no mixing and the ratio is unity) and drops to zero when the mixture is uniform (the root-mean-square fluctuations are zero).

If there was no diffusion, and only the smallest possible eddies were present, the value of I_s would still be unity; thus the intensity of segregation as defined is a good measure of the diffusional process. Equation (4) gives the simplest form of the intensity of segregation (mixing of pure species) and is defined as a function of time-averaged variables at a point. This of course assumes that such an average can be obtained; i.e. that the system is at steady state or is changing slowly when compared with the time necessary to obtain the average. For a complete definition of a given system, one would have to specify the variation of I_s over the entire volume. As a simple example, consider plug flow, in which two fluids are to be mixed. It will be assumed that each fluid is initially distributed uniformly across the pipe cross section on a macroscopic scale under the condition of complete segregation ($I_s = 1$). As the fluid moves down the tube in plug flow, mixing will occur as a result of the turbulent field and diffusion, and the value of I_s will decrease to zero in the limit of molecular uniformity. Actually, I_s must be measured over some small but finite volume. If this volume is too small, submicroscopic variations are detected (statistical fluctuations in the number of molecules present), and if the volume is too large, the measurement becomes insensitive and approaches the average value of the system. For many problems (such as nonideal mixers used for reactors), a detailed study of the variation of I_s over the entire reactor is not desirable, and some space average of the entire system is used. Finally, one can contrast the measurement of I_s with the more conventional "mixing times", which are a direct function of the means of measurement and thus one cannot attach absolute meaning to them.

The definition of the intensity of segregation given in Eq. (4) is restricted to the mixing of two pure species. When recycle occurs, one or both of the species being mixed are contaminated with the other, and a new defining equation is needed. This has been worked out by Rao and Brodkey (1972B) and is given by

$$I_s = a'^2 / (\overline{A} - A_{in\ B})(\overline{B} - B_{in\ A}) \qquad (5)$$

Where $A_{in\ B}$ refers to the amount of A on the average in the B stream. The result is useful where a recycling of streams occurs, such as in mixing vessels where the tracer or reactant is injected below the impeller. The mixing occurs between the tracer and the recirculating stream which contains tracer.

III. EXPERIMENTS AND RESULTS

Experimental results on mixing are limited, but enough results on a variety of geometries are available to allow some understanding of the mixing process and the evaluation, in part, of theoretical approaches. Mixing in the core region of a pipe (centrally located jet injector), in the impeller stream of a continuous flow-stirred tank, and in a multijet reactor configuration have all been measured along with the various turbulence parameters. Two things are desired: one would like to be able to predict the key turbulence parameters from the geometry and, in turn, one would like to predict the mixing from the turbulence parameters. This means that extensive measurements of both the turbulence and concentration fields will have to be made in the same geometry. In this section the various experimental configurations will be described and some typical results will be given. References to more extensive results will be cited.

III.A. *PIPE LINE TURBULENCE AND MIXING*

Extensive measurements have been made of turbulence in pipes, but one precaution must be noted. In mixing experiments, the material to be mixed is fed as a jet along the axis of the pipe. This unfortunately modifies the flow in the region immediately downstream of the injector. The turbulence associated with the wake is greater than for a flow without the injector and the mixing experienced in this region is correspondingly greater. Thus it is best to have turbulence measurements on exactly the same system that is to be used for the mixing.

Brodkey *et al.* (1971A) have reviewed the axial and radial fluctuation data on turbulent pipe flow of fluids and have noted the extent of agreement between workers and the difference between results when gases and liquids are used. Generally, the precision for any given run was considerably higher than the accuracy between runs by different investigators. The deviation between results became more pronounced as the wall region was approached. The liquid results averaged less than gas results, but the differences in the averages were about the same as the deviations in either the air or liquid results. Since the gas flow systems were larger and since hot-wire anemometry was used rather than hot-film or visual methods, it would appear that the gas results are more reliable, since more is known about the heat transfer from hot-wires than from hot-films, and a larger system allows much more accurate positioning and less interference from probe mounts. If it were necessary to select one set of measurements as the most reliable, those of Laufer (1954) are probably the best and are near the average for the gas and liquid results combined.

Turbulence and mixing in a pipe-line system have been investigated by Lee and Brodkey (1963,1964), Brodkey (1966A), Gegner and Brodkey (1966), and Nye and Brodkey (1967A, 1967B). The flow system is shown in Fig. 4 with its

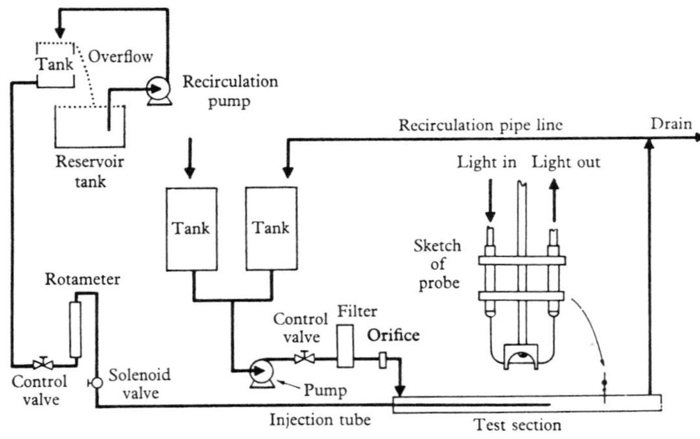

Fig. 4. The pipe flow system.

supply tanks, pump, filter, and 48-ft test section. The injection facility is also shown. The thin stainless steel injection tube is located axially in the test section pipe and dye is injected at the center line. In order to make spectrum measurements, a constant head tank rather than a pump must be used in the injection system, since the basic frequency of the pump might be detected in the spectrum.

Figure 4 also shows the light probe for concentration measurements. It is a fiber optic device in which a high intensity light is shone in at one end; the light is conducted along the fiber optic line to the measuring head where it then jumps across a gap to the other fiber optic line and is taken back along this line to a photomultiplier. Further details of the flow and measuring system are given in the references cited above. Suffice it to say here that such problems as probe size, noise, frequency response, injector interference, etc., all had to be considered. The injector effect is detailed in Lee and Brodkey (1964) and the probe characteristics of an early design in Lee and Brodkey (1963) and of a later model in Nye and Brodkey (1967A).

For the liquid flows of interest here, hot-film anemometry was used for the turbulence measurements. Mean velocity, axial component of the turbulence, and spectra are all needed in order to obtain the descriptive parameters that enter the calculations for the estimation of mixing. A low wave number cutoff k_0 (to be defined later), the microscale, and the turbulent kinetic energy dissipation are all needed and can be obtained from the basic data cited. However, it will be more convenient to introduce these after the theory has been reveiwed.

The basic data obtained from mean and fluctuating concentration measurements involves the distribution of these variables across the pipe radius and along the flow direction. These will not be repeated here as samples can be found in Lee and Brodkey (1964) and in complete detail in Lee (1962). From the basic data, various normalization representations can be used. Again these will not be presented here as they are available elsewhere (Brodkey, 1968; Gegner, 1965; and briefly summarized by Gegner and Brodkey, 1966). Of more importance are the spectra and the decay of the intensity of segregation in the axial direction, for it is these that are directly comparable to the various theoretical efforts. The scalar spectral data as measured by Nye and Brodkey (1967B) are shown in Fig. 5 for three positions downstream from the

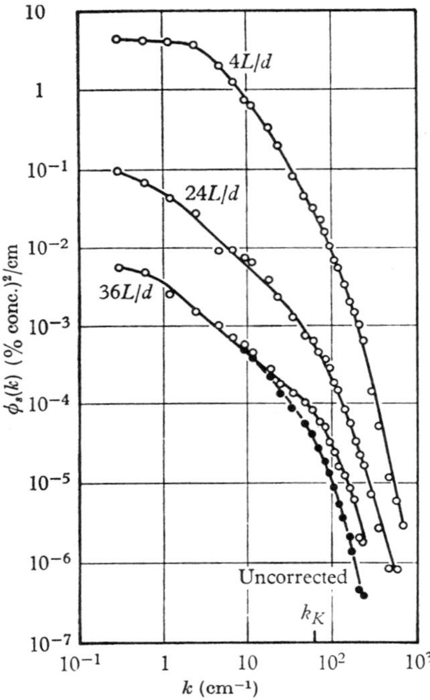

Fig. 5. One-dimensional scalar spectra for pipe flow mixing.

injector. Figure 6, taken from the same work, gives the decay in the intensity of segregation. It is compared to the earlier data by Lee and Brodkey (1964) which was done with a probe of limited resolution. We will return to these and additional pipe flow mixing results when comparisons are made to theory.

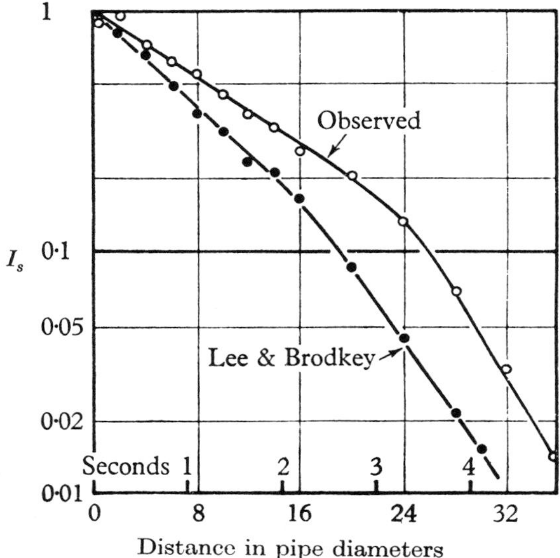

Fig. 6. Decay of I_s along the centerline (observed from Nye and Brodkey).

III.B. *STIRRED TANK TURBULENCE AND MIXING*

Rao and Brodkey (1972A) made measurements of the turbulence in the impeller stream of a continuous flow stirred tank system that is shown in Fig. 7. The impeller configuration is shown in Fig. 8. In the paper, references and comparisons to the work of others were given and will not be repeated here. In pipe flow, the direction of the mean velocity vector is well defined, and the necessary orientation of a Pitot tube or hot-film anemometer follows directly. In a system like a mixing vessel, the velocity vector direction is unknown and must first be established before turbulence measurements can be made and related to the coordinates of the vessel. In addition, the Pitot tube measurements of the magnitude of the velocity vector will be incorrect in very high turbulence regions because of the high turbulence level and the inadequacy of correction methods for this.

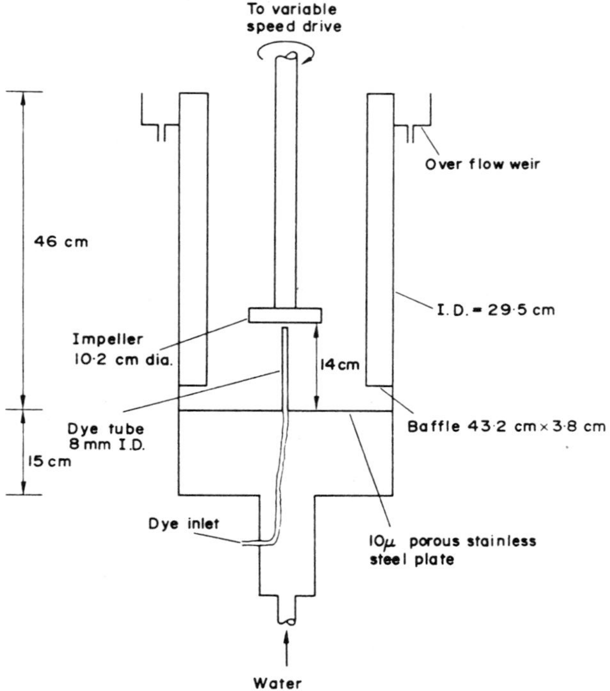

Fig. 7. Continuous flow stirred tank system.

Rao and Brodkey use a multiport null type Pitot tube of standard commercial design to establish the velocity vector direction and uncorrected magnitude. The actual magnitude was determined with an orientated hot-film probe. The relative intensity was fairly uniform being between 48 and 65%, which can be compared to the 3.4% at the center-line in pipe flow. For the open impeller design used, the flow past the hot-film probe was quite intermittent and is a result of the randomly moving jet generated by the impeller blades. Autocorrelations and one-dimensional spectra of the velocity fluctuations were obtained at various positions in the stream. These revealed the presence of periodic velocities close to the impeller generated by the rotating blades. Integral

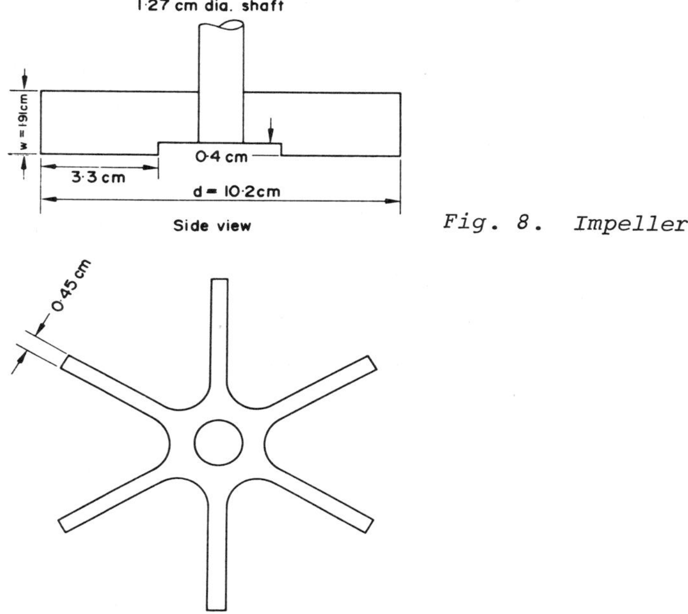

Fig. 8. Impeller

velocity scales were calculated from the autocorrelation after subtracting the contribution of the periodic velocities. Velocity microscales and energy dissipation rates were calculated primarily from the derivative signals of the fluctuating velocities, although other techniques were used for comparison at selected positions. The probability density distributions of the velocity fluctuations were significantly different from the normal distribution and were negatively skewed. Intermittency factors were obtained with an electronic on-off circuit and at a few positions from the flatness factor. The latter method did not yield satisfactory results because of the non-normal distribution of velocities.

The difficulty in making such turbulence measurements is apparent; but, even more difficult were the attempted measurements of the mixing (Rao, 1969). The limited mixing study had to be done as a transient experiment because concentration fluctuations could not be detected at steady state conditions. This was a direct consequence of the recycle effect considered in Eq. (5). At steady state, the effect of the dye component being recycled in the recirculating stream was so close to the average concentration in the inlet stream that the fluctuation $(\overline{A} - A_{in\ B})$ could not be detected. However, during the start of dye injection, the dye in the recirculating stream was small so that measurements could be made at two positions in the stream. The mixing is extremely rapid with I_s being about 0.002 at 1/2-inch from the impeller and 0.0015 at 1-inch. This represents 99.8 percent and better mixed.

III.C. *MULTI-JET REACTOR TURBULENCE AND MIXING*

The turbulence and mixing measurements by McKelvey *et al.* (1974) were done on reactor designs that were identical to those of Vassilatos and Toor (1965) and Mao and Toor (1971). The object was to obtain enough information to relate the mixing with the kinetics as measured by Toor and his co-workers. An overall view of the flow system is provided in Fig. 9. The multi-injection head details are shown in Fig. 10. Details of the system can be found in the references cited. The main difference between the two reactors was the number of injector tubes packed into the 1-inch cross section. In the first model (A), 97 tubes were used on a square array, and in the second model (B), 188 tubes were used on a close-packed arrangement. Besides information on liquid reaction systems, the mixing of gas systems was also studied. This latter has practical importance in gas phase reactions, such as combustion for the supersonic ramjet. Here the question is whether or not the fundamental mixing can be accomplished in the distance provided by the combustion chamber.

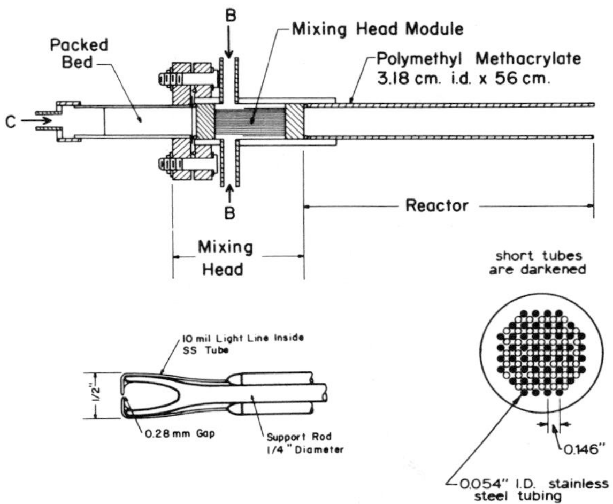

Fig. 9. Multi-jet reactor system and light probe.

On logical grounds, the system should have provided a uniform flow field with a relatively flat velocity profile at the start, slowly changing into the turbulent form as the flow proceeded down the pipe beyond the injectors. The reaction itself would be over before reestablishment of the velocity distribution was accomplished. This is what we thought. But this was not what was found. Instead of a flat profile, there were vorticies along the wall which were associated with the back flow in the region near the wall. This back flow caused considerable change in the mean velocity and all other aspects of the flow. The net result was a complex flow, far more complicated than expected. In Fig. 11, the centerline axial velocity is plotted against axial position, and instead of a uniform flow, there is a very rapid and then a slower decrease. This is associated with a jetting effect which the back flow on the wall induces. Figure 12 shows the intensity of the velocity fluctuations at the centerline as a

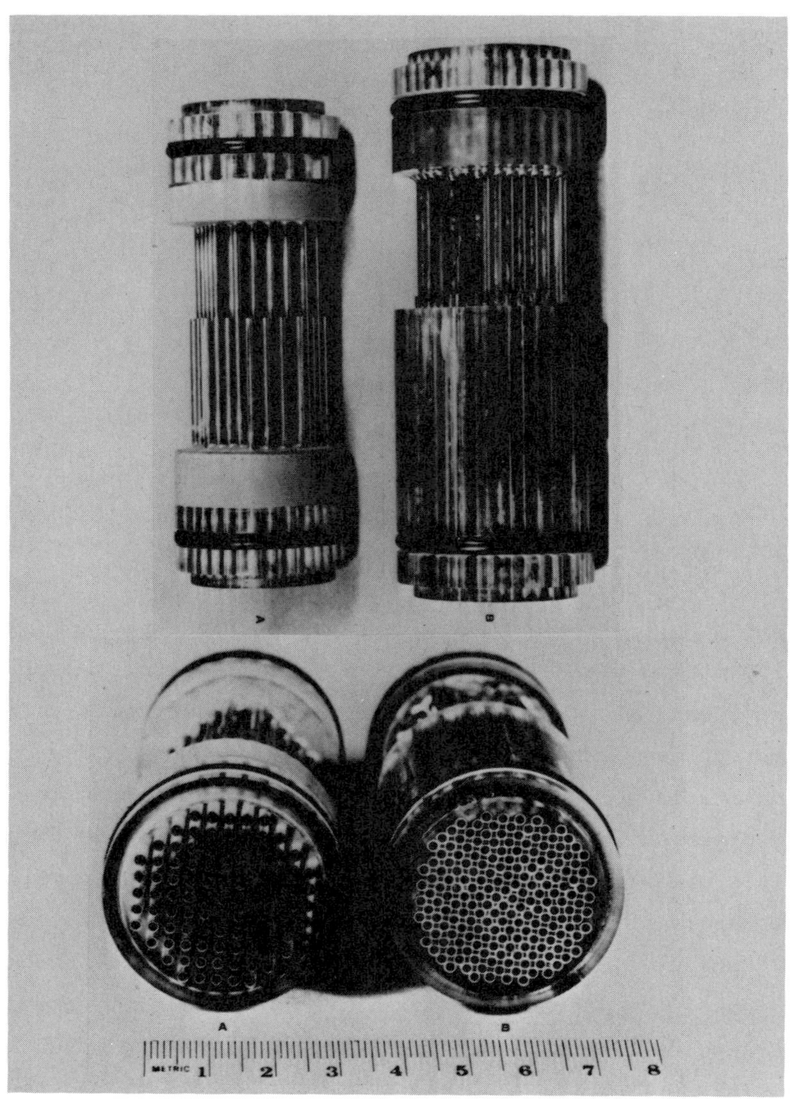

Fig. 10. Injector head details.

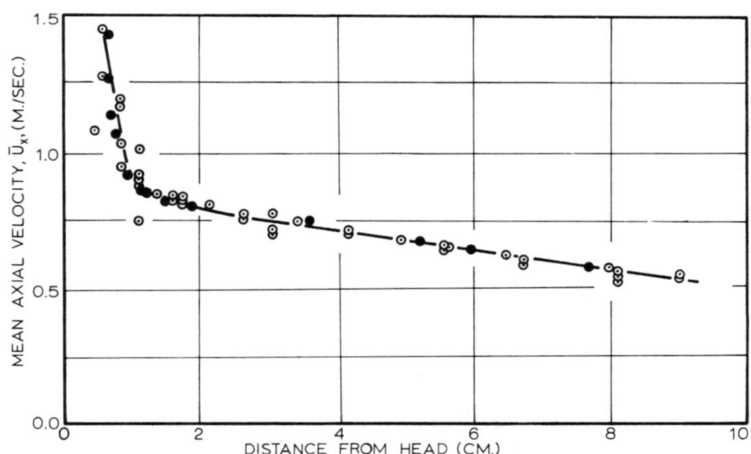

Fig. 11. Average axial velocity along the centerline of the reactor (Model A).

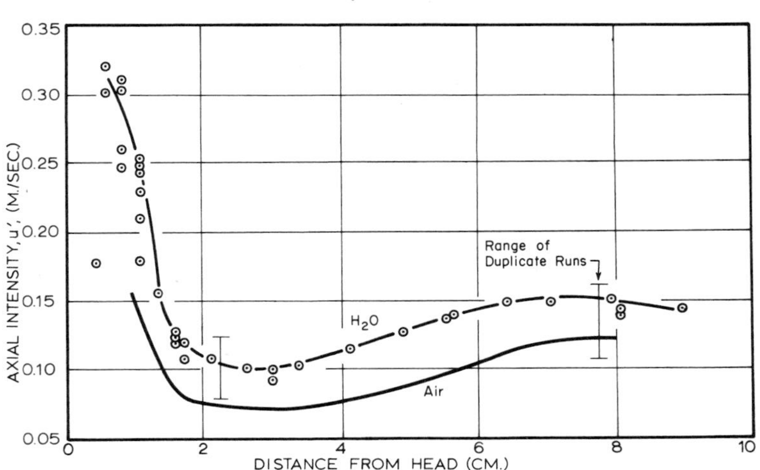

Fig. 12. Axial intensity along the centerline (Model A).

function of axial position. There is a rapid decrease in intensity following the injection from the small tubes; a minimum occurs, and then an increase to an equilibrium value results. The minimum is unusual but is associated with the jet effect. On an identical air system in which we could make more reliable measurements, we have found that the increase in energy was associated with the vortex or a separation that occurs near the wall. Furthermore, we obtained high-speed motion pictures of the flow as a function of axial distance. The difference in the appearance of the field was dramatic at the region of minimum axial intensity and the region far from the head.

The modification to increase the number of tubes was helpful but did not eliminate the nonideal gross flow characteristics of the system. However, this was not critical for the study since the flow patterns in the reactors were the same when Toor and his co-workers measured the reaction conversions and when we measured the turbulence and mixing.

Detailed turbulence measurements were made for both water and air flow for the 97 injector model (A) and for air only for the 188 injector unit. If properly scaled, the mean velocity along the centerline as a function of axial position was identical for the water and air flows at the same Reynolds number. Thus, the measurements in air for the closer packed reactor could be used for liquid flows. This is important for later comparisons to the reaction results obtained by Mao and Toor. Autocorrelation, spectra, and probability density distribution results were obtained from digitized data as described by Brodkey *et al.* (1971A). These in turn were used to compute the characteristic turbulence parameters: velocity macroscale and microscale, low wave number cut off (k_0), and the kinetic energy dissipation. As mentioned earlier, these are essential for estimating the mixing from a knowledge of the turbulent field. Interpretation of the spectra in terms of the flow field, detailed radial traverse studies, turbulent kinetic energy and Reynolds stress variation, probability distributions, etc., can all be found in McKelvey *et al.*(1974).

A slightly modified light probe system was used to measure the concentration field. Details necessary for analyzing concentration signals with regard to linearity and intensity are given by McKelvey (1968). Because of the extremely efficient mixing of the reactor module, the concentration fluctuations were always a small percentage of the average concentration. However, because of the low levels, noise problems were far more severe in the present application than in previous uses of the probe. Although satisfactory results were obtained using the light probe, newer developments in the use of conductivity measuring devices (Torrest and Ranz, 1969) should be considered for future use.

The most important single statistic of the scalar field is the variation of $I_s = a'^2/a_0'^2$ which is shown in Fig. 13. Downstream of the coalescence plane, the decay is well represented by a -3/2 power decay law. This dependency had previously been observed by Gibson and Schwarz (1963) and by Keeler *et al.*(1965) and predicted theoretically by Hinze (1959). All of these results, however, were obtained in velocity fields which were approximately isotropic, and it is somewhat surprising that the same decay is observed in such an anistropic field. It may be, therefore, that the decay of passive scalar fields is rather insensitive to many of the details of the decay of the velocity field, and that scalar fields associated with decaying anisotropic velocity fields

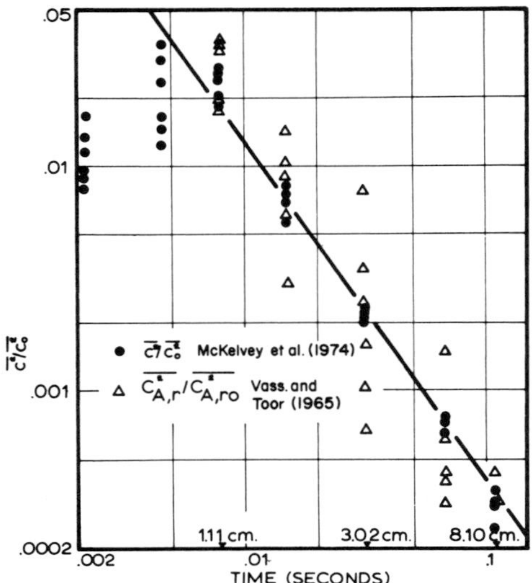

Fig. 13. Decay of the concentration fluctuations (Model A, ordinate is I_s, $C_A = A$, $c = a$).

approximate the -3/2 law.

In spite of the large amount of background noise in the concentration signal, the one-dimensional scalar spectra were obtained (see Fig. 14); unfortunately, much of the spectra lies beyond the cutoff frequency of the light probe and could not be obtained. However, when far enough down-stream of the ejectors, a -1 power region was observed which has been suggested by Batchelor (1959) and was mentioned earlier during the discussion of the pipe flow mixing work by Nye and Brodkey (1967B). The various parameters associated with the scalar field were also determined by McKelvey *et al.* (1974).

Further references on methods etc. can be found in the outline given in the appendix.

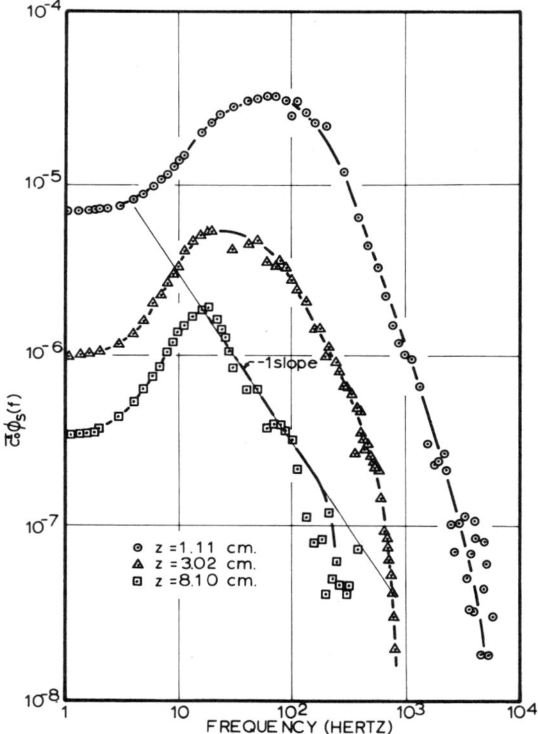

Fig. 14. One-dimensional scalar spectra (c = a, z is distance from reactor head).

IV. THEORY

There have been several simplified approaches to the prediction of mixing. These have been valuable because they can provide some insight into the relationship between turbulence and mixing and can suggest to us some of the controlling variables that should be considered in commercial mixing operations. Under somewhat idealized conditions, such as in the experiments previously described, the theories allow us to estimate the time required for a specific degree of mixing. Under such conditions, one can even, at times, estimate the extent of chemical reaction that will occur in homogeneous reactions involving two separate feeds and in which the degree of turbulence is a contributing factor.

In this section, the details of the theory will not be given, but rather an idea as to the approach used will be presented. We shall emphasize the results and the implications that these have for mixing operations. But before we can start on the mixing per se, some background about turbulence is needed.

IV.A. *TURBULENCE*

Our understanding of turbulent mixing is hampered by our lack of knowledge about the dynamics of turbulent motion. There are two distinct modes of approach to the motion problem. One is phenomenological theories such as the Prandtl mixing-length theory. Even though this appraoch provides an over-all solution with a modest degree of accuracy for practical problems, it is strictly founded on physical intuition and is generally recognized to be limited in its possibilities for further development.

The other approach is the statistical theory based on the random behavior of eddies in a turbulent field. This theory of turbulent motion has attracted many capable theorists and experimentalists since Taylor (1938) initiated the notion of a statistical approach. At the present stage of development, however, this approach is still far from complete for use on practical design problems. This theory also presents many difficulties arising from the indeterminateness of the equation which represents the basic law of momentum transfer. It should be strongly emphasized that the description of turbulent mixing always includes the unknowns from turbulent motion which must be understood before one can solve the turbulent mixing problem. In short, there is so far no clear-cut, determinate equation or system of equations for turbulent motion.

The mechanism of turbulent motion is of such a complex nature that at present we are unable to formulate a general physical model on which to base an analysis. Thus, we approach the problem from a rigorous statistical theory in which certain simplifying assumptions can be introduced that will allow us to solve for some of the variables of interest. One must view the subject as a reasonably rigorous mathematical representation with some simple and limited mechanistic ideas injected. We simply do not know for certain the details of what is physically occurring in turbulence, and thus we are unable to express the picture in mathematical terms. There is, however, considerable work recently reported in the literature and currently underway that is directed towards the understanding of the coherent structures that appear in turbulent shear flows (see for example Corino and Brodkey, 1969; Wallace *et al.*, 1972; and Nychas *et al.*, 1973, references therein and the appendix). As more and more is learned about the mechanism of turbulent shear flow, we realize more and more the inadequacies of the early statistical approach and the necessity of eventual modification of it.

To proceed, we will asssume the reader has some familiarity with the statistical theory of turbulence; if not, the chapter on TURBULENCE AND MIXING in Brodkey (1967) will serve as a background reference for this. What we need here is the means for evaluation of the various parameters of statistical turbulence from the basic data cited earlier. By the parameters of statistical turbulence, we mean those descriptive terms obtained by processing the instantaneous signal beyond just the rms value. For the most part, such analysis has been restricted to turbulence in the axial direction (u_x). There have been measurements in other than the axial direction, but these are much harder to obtain. For the background details one can see Brodkey (1967), the references therein, and the appropriate references in the extended outline provided in the appendix.

The main parameters of importance in mixing that are obtained from the autocorrelation and spectrum are microscale, macroscale, low wave number cut-off and turbulent energy dissipation. These can usually be determined in more than one way, thus offering an internal check on precision of the calculations.

The *microscale* or sometimes called *dissipation length* is defined as

$$\lambda^2 = 1/(d^2f/dr_x^2)\Big|_{r_x = 0} \tag{6}$$

where f is the velocity autocorrelation. Equivalent to this is

$$\lambda^2 = u_x'^2/\overline{(\partial u_x/\partial r)^2} \tag{7}$$

with

$$\partial u_x/\partial r = (1/\overline{U}_x)(\partial u_x/\partial t) \tag{8}$$

which is Taylor's hypothesis. The r.m.s. velocity fluctuation in the flow direction is u_x'. Another equivalent form is

$$\lambda^2 = u_x'^2/2\int_0^\infty k^2\phi(k)dk \tag{9}$$

where k is the wave number and ϕ is the one-dimensional spectrum. Still another method involves the assumption of a normally distributed variable

$$\lambda^2 = \overline{U}_x^2/\pi^2N(0)^2 \tag{10}$$

where N(0) is the average rate of sign change or the density of zero-crossings. The best values of the microscales and other parameters for the various flows being considered will be summarized later when some rather crude semi-empirical estimates for them have been introduced and comparisons can be made.

The *macroscale* can be difficult to determine from autocorrelation data. It is defined as

$$L_f = \int_0^{\infty} f(r_x)\, dr_x \qquad (11)$$

Although the macroscale is strictly defined by Eq. (11), its primary value is a measure of the *large scale* fluid motions. In general, most turbulent fields which have been examined experimentally do not exhibit strong periodic motions and $f(r_x)$ has generally been positive. But in some flows $f(r_x)$ can oscillate between positive and negative, and the net value from Eq. (11) can be very small, even smaller than the microscale. Thus, in many respects, the area under the correlation function from $r_x = 0$ to the point at which the correlation first becomes negative is a better measure of this scale, or even better would be an integration using $|f(r_x)|$. For many systems a more reliable estimate can be obtained from the relation

$$L_f = (\pi/u_x'^2)\ \phi(0) \qquad (12)$$

where $\phi(0)$ is the one-dimensional spectrum at zero wave number.

The *dissipation* of the *energy* of *turbulence* can be calculated using the isotropic relationship

$$\varepsilon = 15\nu u_x'^2/\lambda^2 \qquad (13)$$

and the best value of λ. Although the turbulence field in mixing operations is usually not isotropic, no other relationship is available, and hence, Eq. (13) serves as an approximation.

The term k_0 is a low wave number cut-off point that is defined so as to retain known relations between u'_x, E(k) and ε with a defined form for E(k) (see McKelvey, 1968). It is

$$k_0 = (2/3)^{3/2} \varepsilon/u'^3_x \tag{14}$$

In general there are equivalent parameters for the scalar field to those defined above.

The various scales, microscales, wave numbers, and dissipation parameters would be most useful if they could be estimated without measurement of autocorrelation or spectrum, as it is these that appear as the controlling parameters of mixing. Let us assume the only information available is the system (fundamental properties such as ν), geometry (L), and the r.m.s. fluctuation, u'_x. The following have been suggested as reasonable but somewhat crude estimates:

$$k_0 = 2/L \tag{15}$$

$$L_f = (3/4)(1/k_0) \tag{16}$$

$$\lambda^2 = 10\ \nu L_f/1.1\ u'_x \tag{17}$$

where L is a characteristic length; for pipe flow this is the radius. Combining gives

$$\lambda^2 = 3.41 \nu L/u'_x \tag{18}$$

Combining Eqs. (13) and (18) gives

$$\varepsilon = (15/3.41)(u'^3_x/L) = 4.40\ u'^3_x/L \tag{19}$$

A slightly alternate relation for ε can be obtained by combining Eqs. (14) and (15) directly:

$$\varepsilon = 3.68\ u'^3_x/L \tag{20}$$

The difference is minor when one considers the gross approximations involved.

A comparison between the experimental values of the characteristic parameters for the various flows and the semi-empirical estimates above will be reserved for a later section.

IV.B. *MIXING*

The *dispersion* of a contaminant by turbulent motion is of fundamental importance in many problems. To illustrate, the oceanographer would like to be able to determine beforehand the dispersion of wastes discharged into the sea; the degree of air pollution and the proper design of stacks depend on the ability of the turbulent wind to disperse smoke; and the time necessary for blending will depend on one's ability to disperse the material to be mixed. Such dispersions are often called *turbulent diffusion*; the analysis of such problems for the most part are concerned with the case of no superimposed molecular diffusion, and, in effect, can be considered as the motion of marked fluid particles or elements.

A Lagrangian view can be used to gain some insight into the mechanism. Consider individual elements that leave a fixed point in space: first, for the case of no mean motion, the various elements will be carried from the source by the turbulent eddies. Those caught in a large eddy (with generally large motion) would be expected to be carried further than those that are initially a part of a small eddy. Thus, at any instant in time, there will be a distribution of elements about the point source. This may be easier to visualize by superimposing a uniform mean velocity on the turbulent field. Each element or particle leaving the point source would be expected to deviate from the linear path in a random manner depending on the local nature of the turbulence. The r.m.s. deviation for the particles would be observed as a continued divergence, spread, or dispersion as the particles are carried downstream from the point source by the uniform mean velocity. This is an eddy motion and can occur in the absence of molecular diffusion. In this discussion, the distance from the point source divided by the uniform mean velocity has been used to replace time in the first illustration. It was Taylor (1921) who considered the diffusion of infinitesimal fluid particles from a point source in a homogeneous isotropic field with no superimposed molecular diffusion. The theory was restricted to eddy diffusion from a source in a static field or from a source moving with a uniform velocity field.

What we shall call the *turbulent mixing problem* is more complicated. We are interested not only in the spread of the material to be moved, but in addition, how it becomes essentially homogeneous with the surrounding fluid. For this we need molecular diffusion. Because of the nature of the problem, little is known about the actual mechanism. Consequently, an approach similar to statistical turbulence is used; i.e., the problem is formulated rigorously in terms of statistical averages without references to any specific model. There are two important aspects of the problem. First, experimental information interpreted in terms of the theory may provide some insight into the actual mechanistic contribution of turbulence and of molecular diffusion to mixing; and second, with reasonable approximations for the mixing spectrum and boundary conditions, the theory can be used to predict the time of mixing under specific mixing conditions. Admittedly, we have not progressed in either direction as far as we would like; however, as will be seen, progress has been made.

The statistical theory of turbulent mixing has been

developed parallel to the turbulent motion problem. The basic equation for turbulent mixing is that of mass (or heat) conservation which is the counterpart of the nonlinear Navier-Stokes equation for turbulent motion. It is obvious that the treatment of mass (or heat), which we shall call *scalar quantity*, is much simpler than that of turbulent velocity which is a vector quantity. However, the problem of turbulent mixing of a scalar quantity enjoys all the difficulties that turbulent motion does because of the nonlinearity of the governing physical equations. The velocity fluctuations are always introduced as a part of the unknown functions in the course of describing turbulent mixing in a turbulent field. Therefore, the evaluation of the various functions for turbulent motion, such as correlation and spectrum functions, must be available before one attempts to solve the turbulent mixing problem. The closure of the infinite set of moment equations in mixing is analogous to that of the motion problem. If the analysis of turbulent mixing in an isotropic homogeneous turbulent field (the most idealized physical reality) can be applied to real shear problems, one can arrive at some sort of approximation to the mixing problem. If local isotropic turbulent conditions exist, then the approximation may be good; however, few experiments to indicate the degree of approach to local isotropy have been reported. Very little theoretical or experimental work has been done on turbulent mixing of two initially separated feed streams entering a reactor.

The procedure for obtaining the equation descriptive of turbulent mixing is given by Corrsin (1951) and outlined by Brodkey (1967). It should be emphasized that it is based on nothing more than the conservation of matter and the assumption that the material can be considered as a continuum.

For isotropic conditions the basic differential equation, when converted to wave number space, is

$$\partial E_s(k)/\partial t = T_s(k) - 2Dk^2 E_s(k) \tag{21}$$

where $E_s(k)$ is the scalar spectrum. Integration over all wave numbers of the spectrum is related to the intensity of segregation by the relation

$$I_s = a'^2/a_0'^2 = (1/a_0'^2)\int_0^\infty E_s(k)\, dk \tag{22}$$

Certain intuitive ideas have been developed which suggest the form that the spectrum should take. Figure 15 shows the

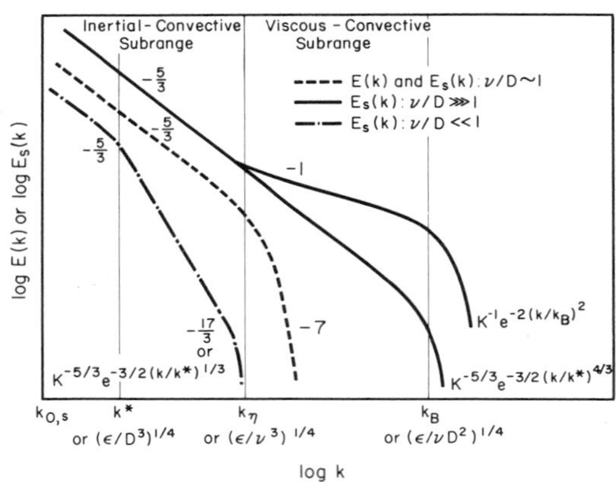

Fig. 15. Intuitive and phenomenological theoretical spectra.

intuitive and phenomenological results obtained previously on the velocity spectrum and from somewhat similar analyses for the scalar field. For example, in the viscous-convective subrange, owing to the great difference in ν and **D**, $(\nu/D >>1)$, the spectrum of concentration fluctuations extends much further into the high wave number range than that of the velocity fluctuations. One would expect a range of $E_s(k)$

(conc^2-cm), which would depend on ε_s (conc^2/sec), the effect of the large velocity eddies, stretching out the scalar blob [a strain rate parameter $\gamma(\text{sec}^{-1})$], and on $k(\text{cm}^{-1})$. But $E_s(k)$ would not depend on the diffusivity since the diffusivity would be important beyond the high wave number end of this subrange. Dimensionally, one must have

$$E_s(k) \propto (\varepsilon_s/\gamma)k^{-1} \tag{23}$$

which is the form derived by Batchelor (1959) by both dimensional and analytical arguments. In Fig. 3 some of the concentration eddies look contorted (A) and some look elongated (B). The blob marked A is of the general shape one would expect from the action of eddies smaller than the blob (i.e., a situation that might exist in the inertial-convective subrange, -5/3). The blob marked B has the shape one would expect as a result of a pulling action exerted by large eddies of large velocity on a smaller blob (i.e., a situation that might exist in the viscous-convective subrange, -1). One might picture this as material caught between two large rollers. We will make further use of these intuitive guesses when we treat the details of the turbulent mixing problem. We will digress here to note that the -1 range has been observed in the mixing experiments previously described (Figs. 5 and 14) by Nye and Brodkey (1967B), McKelvey *et al.*(1974) and Grant *et al.* (1968).

The *theory of Beek and Miller* (1959) based on a suggestion by Corrsin (1951) was one of the first analyses of the scalar mixing problem. The analysis was based on the idea of closing Eq. (21) with an eddy-diffusivity type of transfer function that paralleled an earlier suggestion by Heisenberg for the velocity field. This involves the use of a scalar spectrum for both liquids and gases that has -5/3 and -7 regions. Although not correct for liquid systems because the true spectrum follows the -1 rather than -7 assumed (See Fig. 15), it is interesting because, as it turns out, much of the details of the spectrum are lost since the intensity of segregation or degree of mixing depends on an integration of the spectrum curve and not on the curve itself [Eq. (22)]. As is well known, integration is a smoothing process. Beek and Miller integrated the basic equation under certain idealized assumptions to obtain the intensity of segregation as a function of distance down an idealized pipe line in which there was no velocity distribution (plug flow). The results are quite logical even though they may not be correct in magnitude because of the assumed spectrum. In Fig. 16,

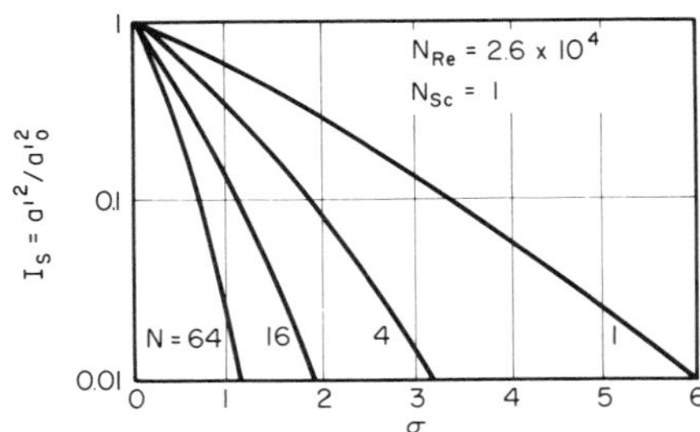

Fig. 16. Decay of I_s for Gases (N is number of injectors).

mixing of gas in a pipe line is considered. The term, σ is a dimensionless measure of the distance down the pipe, while N can be pictured as the number of injectors. For the mixing of gas in a pipe line, increasing the number of injection points is associated with a reduction in scale which in turn provides a larger area for rapid diffusion associated with gas systems. The results for liquid systems (Fig. 17) show a modest dependency on the number of injectors. Here, the diffusivity is so low that the turbulence has enough time to distribute the clumps of material into their smallest eddy sizes before much diffusion has occurred. Thus, one would

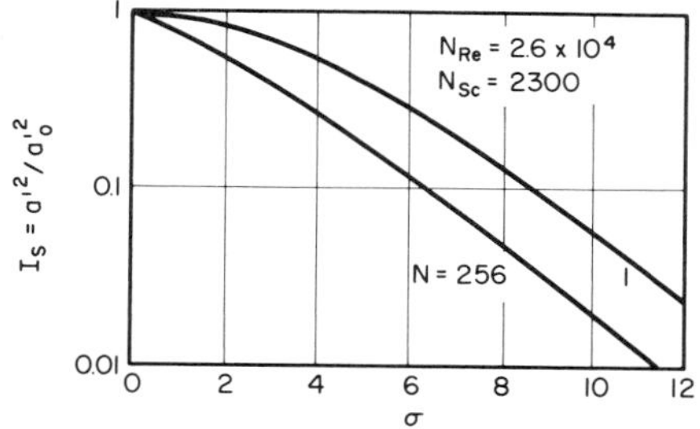

Fig. 17. Decay of I_s for Liquids (N is number of injectors).

expect that liquid mixing would be less dependent on initial conditions or number of injectors. The time of mixing for both the liquid and gas system is highly dependent upon the Reynolds number. The decrease in mixing time for the gas just compensates for the increase in velocity associated with the higher Reynolds number. For liquids, the effect of the increase in turbulence more than off-sets the increase in velocity, thus resulting in a net decrease in the length of pipe required. This latter result is open to question from similarity considerations.

It has often been suggested that the Reynolds number be used for scale-up of pipeline mixers (see Brodkey, 1967). We may question the validity of this and base our analysis on the following considerations. For constant Reynolds number, $u_x'/\overline{U}_x$ is constant. Since $\overline{U}_x$ must decrease in order to offset the increase in d_0 for the same Reynolds number, u_x' must also decrease. Furthermore, by Eq. (15), if d_0 increases, k_0 must decrease. From the expression for σ, $\sigma = k_0 u_x' t$, one sees that t must increase considerably for a given σ with an increase in d_0 (actually with d_0^2). However, since the time can be taken as length / $\overline{U}_x$, t increases only with d_0. Therefore, to obtain the same degree of mixing, one must also increase the length in proportion to the increase in d_0. This is, in effect, a requirement of geometric similarity. Even scale-up on the same velocity will not be sufficient, since the increase in Reynolds number will not change $u'_x/\overline{U}_x$ appreciably, which must offset the decrease in k_0. Thus, in all cases, to obtain the same degree of mixing, one must also design for an even greater length than in the smaller test section. Toor and Singh (1973) have recently considered scale-up in much more detail. The results are the same for the constant Reynolds number case and are developed in more detail for the variable Reynolds number case.

The conservation of matter relation, which was used to obtain Eq. (21), can be reduced to

$$da'^2/dt = -12Da'^2/\lambda_s^2 \tag{24}$$

where λ_s is the scalar microscale and is defined in terms of the correlation of Eq. (2) in a manner analogous to Eq. (6). If one can assume the microscale does not vary drastically, the equation can be integrated to give an expression for the intensity of segregation:

$$I_s = e^{-t/\tau} \tag{25}$$

where

$$\tau = \lambda_s^2/12D \tag{26}$$

and is the time constant of mixing. This has been suggested by *Corrsin (1957)*, who has made estimates for the mixing time constant. Corrsin (1964A) has used the spectrum estimates given in Fig. 15 to estimate λ_s and thus τ for both the low and high Schmidt-number ranges. For the low Schmidt number case integration of the curve from $k_{0,s}$ to k_η gives, after simplification,

$$\begin{aligned} \tau = \lambda_s^2/12D &= 2/(3-N_{Sc}^2)(\varepsilon k_{0,s}^2)^{1/3} \\ &= (5/\pi)^{2/3}[2/(3-N_{Sc}^2)](L_s^2/\varepsilon)^{1/3} \end{aligned} \tag{27}$$

where L_s is the scalar macroscale defined by Eq. (3), ε is the velocity energy dissipation per unit mass given by Eq. (13) or (19) and $k_{0,s}$ is the wave number representative of the large scalar blobs, which would be analogous to $\mathbf{k}_0$ of Eq. (14), associated with the energy-containing eddies. It is defined so as to retain known relations between a', $E_s(k)$, and ε_s with a defined form for $E_s(k)$ and is

$$k_{0,s}^{2/3} = (3/2)\varepsilon_s/a'\varepsilon^{1/3} \tag{28}$$

The term $k_{0,s}$ has been eliminated from the last part of Eq. (27) by the approximate relation

$$k_{0,s} = (\pi/5)L_s^{-1} \tag{29}$$

Corrsin has also shown that the mixing time constant can be expressed as

$$\tau = (\lambda^2/10\nu)(k_0/k_{0,s})^{2/3}[2/(3-N_{Sc}^2)] \tag{30}$$

For the high-Schmidt-number case, the time constant can be obtained by integration of the spectrum curve from $k_{0,s}$ to give

$$\begin{aligned} \tau &= (1/2)[\{3/(\varepsilon k_{0,s}^2)^{1/3}\} + (\nu/\varepsilon)^{1/2} \ln N_{Sc}] \qquad (31) \\ &= (1/2)[3(5/\pi)^{2/3}(L_s^2/\varepsilon)^{1/3} + (\nu/\varepsilon)^{1/2} \ln N_{Sc}] \end{aligned}$$

Note that this equation gives the proper limit of no mixing in the limit of infinite Schmidt number ($\tau = \infty$, which gives $I_s = 1$ for all time).

In many cases one is only interested in scale-up of mixing units. The following is a brief summary of Corrsin's results for this practical problem. In the equations for the time constant of mixing, the most important term is

$$L_s^2/\varepsilon$$

The turbulent energy dissipation can be represented by the

power input per unit mass and an efficiency:

$$\varepsilon = \eta P/M \tag{32}$$

where η is the efficiency of turbulent production, P is the power and M is the fluid mass. Combining gives the scale-up as

$$\eta' P'/M' L_s'^2 = \eta P/M L_s^2 \tag{33}$$

The mass scales with the geometry or K^3 and the mixing scales as K_s. If the efficieny varies with the scale-up, then the result for the power is

$$P' = K_\eta K^3 K_s^2 P \tag{34}$$

If efficiency is constant, the scale-up is

$$P' = K^3 K_s^2 P \tag{35}$$

Finally, if both the mass and the scalar scale-up are the same, the scale-up would be

$$P' = K^5 P \tag{36}$$

While K depends on geometry, K_s would be expected to depend more on the nature of the injection of the material to be mixed. The fifth power can be derived also from pure blending relations (i.e. the number of tank turn-overs held constant on the scale-up); however, the empirical value is nearer to the fourth than the fifth power. From the equations above, one would expect this to be between the third and fifth power.

Rosensweig (1964,1966) derived an expression for the decay of the scalar variance for an isotropic turbulent field and a homogeneous scalar field in a mixing vessel. Beginning with the mass balance equation and using volume averaging of the concentration terms, it can be shown that for a mixer with input and output streams, and its contents uniformly homogeneous, the scalar decay is related by

$$da'^2/dt - \Sigma(\overline{A}_i - \overline{A})^2/\tau_i + (a'^2/\tau) = -\varepsilon_s \qquad (37)$$

where τ_i and $\overline{A}_i$ are the residence time and the average concentration for the ith inlet stream respectively, τ is the average residence time for the vessel, $\overline{A}$ is the average concentration in the vessel and ε_s is the scalar dissipation. In Rosenweig's work he used certain statistical turbulence relationships to relate ε_s to other system variables. As shown by Brodkey (1967) and by Rao and Brodkey (1972B), his analysis was restricted to gas systems, but this is, of course, not necessary. Brodkey (1967) showed that in terms of the intensity of segregation, Eq. (37) can be expressed as

$$I_s = 1/[1 + (\tau_r/\tau)] \qquad (38)$$

where τ is the time constant of mixing and τ_r is the residence time in the vessel. Equation (38) can be contrasted to Corrsin's result given by Eq. (25). Rao and Brodkey (1972B) further showed by an example that Rosenweig's analysis is valid for a well stirred mixer whose contents have a uniform variance and Corrsin's for the continuous flow case. To predict the local variance in a mixer with a non-uniform variance, one would have to divide the mixer into a large number of segments, each of which may be considered as well mixed. Such an approach has been taken by Patterson and will be discussed in Chapt. V.

Finally, consideration must be given to the estimation of mixing with the information provided in the foregoing. Comparison of Eqs. (27) and (30) gives

$$(\pi/5)^{2/3}(L_s^2/\varepsilon)^{1/2} = 1/(\varepsilon k_{0,s}^2)^{1/3} \qquad (39)$$

$$= (\lambda^2/10\nu)(k_0/k_{0,s})^{2/3}$$

which is valid for gas systems, but will be assumed to be generally valid since the gross features of the scalar and velocity fields should depend on the manner of injection and not on the Schmidt number level. Next, it is assumed that $k_0 = k_{0,s}$ since the concern is with cases where the scalar material is grossly dispersed across the geometry. Thus, with Eq. (18), Eq. (39) becomes

$$(\pi/5)^{2/3}(L_s^2/\varepsilon)^{1/2} = 1/(\varepsilon k_{0,s}^2)^{1/3} \quad (40)$$
$$= (\lambda^2/10\nu) = 0.341\ L/u_x'$$

An equivalent relation is

$$(\pi/5)^{2/3}(L_s^2/\varepsilon)^{1/2} = 1/(\varepsilon k_{0,s}^2)^{1/3} = (3/2)(u_x'^2/\varepsilon) \quad (41)$$

This estimate together with ε estimated from Eq. (19) can be used in Eqs. (27) and (31) to estimate the time constant of mixing. If L_s is desired, it can be obtained from Eq. (40). The approximate relation

$$L_s = (\pi/5)(1/k_{0,s}) \quad (42)$$

can in turn be used for $k_{0,s}$. From all of this, one only need know L, ν and u_x' to make an estimate of the mixing. The Reynolds number dependence of mixing enters through u_x' and in the next section comparisons of these will be given with the experimental results.

One further aspect of the theory that must be reviewed is associated with very-fast chemical reactions. The fraction conversion has a simple relation to the degree of mixing or I_s if the reaction is rapid irreversible, second order, between dilute materials with equal diffusivities and in stoichiometric ratio. Toor (1962) showed that if the mixture were stoichiometric, the fractional conversion would be equal to the degree of mixing, that is, $1 - \sqrt{I_s}$. We can easily see the equivalence if we remember that the final mixing to a completely homogeneous system involves diffusion. For simple mixing, each molecule that diffuses from the high- to low-concentration area has a double reduction effect on the driving force by simultaneously reducing the high concentration and increasing the low. Of course, the diffusion is equal molal, and the solvent is also transferred. Now consider the rapid irreversible reaction in stoichiometric balance; the scalar material diffuses into the other material and reacts, and by equal molal counter-diffusion, the reverse occurs and also reacts. The net double effect is still present; however, the double reduction now occurs in the scalar field. On the average, one quantity of scalar material has moved out and reacted, and one has been removed by reaction within the field. Thus, for this special case, the reduction in diffusional driving force is the same and the conversion is equivalent to the degree of mixing. For nonstoichiometric mixtures, Toor obtained an estimate for the increase in conversion as a function of the square root of the intensity of segregation by assuming a normal distribution of the concentration fluctuations about its mean. Toor concluded that "increasing the concentration of one species...is most effective when a high conversion is required and the mixture is initially close to stoichiometric". These results can be used as an extension to any theory of the mixing of nonreacting materials. Some verification of this theory is available. Keeler *et al.* (1965) measured the turbulent field behind a grid and were able to predict the conversion for an ammonium hydroxide-acetic acid reaction. Vassilatos and Toor (1965) obtained similar results for four different neutralization reactions (ionic) in a specially designed jet reactor. In their case the actual turbulent field was unknown so that $1 - \sqrt{I_s}$ was established from the kinetic experiments at stoichiometric ratio. Further direct confirmation will be given in the next section.

V. EXPERIMENTS AND COMPARISONS TO THEORY

Comparisons of experimental data with the theories can be made for both the turbulent and scalar fields. The easiest manner in which to proceed is to treat each geometry in turn and see how far the comparison can be carried out.

V.A. *PIPE FLOWS*

For the turbulent field, one would like to compare the turbulence parameters estimated from the fluctuation and corresponding spectrum data with the rough estimates that can be made using only the fluctuation level and a characteristic length. The results are summarized in Table 1. Those for the water in the 3-inch line were done in the same facility as were the mixing experiments to be discussed shortly. Those for trichloroethylene in the 2-inch line were obtained in the visual flow apparatus described by Corino and Brodkey (1969) and Brodkey *et al.* (1971B). In Table 1 are given the best experimental values along with the equations used to obtain them. For comparison, the calculated rough estimates are given. As indicated in the theory, alternate means are available to obtain the best experimental values. Intercomparisons of these can be found in McKelvey *et al.* (1974) and in the thesis cited therein.

Inspection of Table 1 shows a generally satisfactory agreement between the best experimental values and the rough estimate values. The agreement is certainly satisfactory for the eventual use in making estimates of the time constant for mixing. As to the experimental values themselves, the variation for the trichloroethylene flow with

TABLE 1 *Comparisons for Pipe Flows*

	Equation #		H_2O		Trichloroethylene					
			$r/r_0 = 0$		$r/r_0 = 0$		$r/r_0 = 0.35$		$r/r_0 = 0.75$	
	exp.	est.	exp.	est.	exp.	est.	exp.	est.	exp.	est.
r_0 cm			3.865		2.505					
ν $\frac{cm^2}{sec}$			0.01		0.00374					
u'_x $\frac{cm}{sec}$			2.325		1.35		1.81		2.52	
$N_{Re} = d_0\bar{U}_x/\nu$			40,600		40,600					
$k_0 cm^{-1}$	(14)	(15)	0.70	0.5175	0.86	0.762	0.94	0.762	1.07	0.762
L_f cm	(12)	(16)	1.42	1.45	1.21	0.984	0.88	0.984	0.54	0.984
λ cm	(9)	(18)	0.223	0.236	0.162	0.158	0.144	0.146	0.119	0.129
ε $\frac{cm^2}{sec^3}$	(13)	(19)	16.2	14.3	3.9	3.9	6.8	6.6	16.1	13.9
$N_{Re,\lambda} = \lambda u'_x/\nu$			51.9	54.9	58.4	56.9	51.9	52.6	42.9	46.5
$k_\eta = (\varepsilon/\nu^3)^{1/4}$ cm^{-1}			63.5	61.2	94	94	108	107	134	130

radial position is consistent with our picture of pipe flow; i.e., the scales increase towards the centerline (representative wave number decrease) and the turbulent dissipation decreases. Note that for the macroscale and its corresponding wave number, the rough estimate equations do not allow for this variation since the size is only a function of the pipe diameter.

Next we can turn to the mixing experiments. Enough experimental data have been obtained in the turbulent core region of pipe flow so that comparisons can be made between these experiments and the previously cited theories. The details are given in Gegner and Brodkey (1966), Brodkey (1966A) and Nye and Brodkey (1967B). The earlier data, which are known not to be perfect because of probe resolution, would still show correct tendencies and can be compared to the theoretical predictions. In Fig. 18, a comparison is made with a prediction of Corrsin's analysis. As can be seen, the estimation is quite good.

First, let us describe how the prediction was made (Brodkey, 1966A). The time constant of the mixing comes from Eq. (31). The first term is estimated from Eq. (40) and ε is obtained from Eq. (19) to allow determination of the second term. This, then, is all that is needed to obtain the solid curve in Fig. 18 by the use of Eq. (25); i.e.,

$$\tau = (1/2)[3(0.567) + (0.01/14.3)^{1/2} \ln 3850] = 0.96 \text{ sec}$$

where

$$N_{Sc} = \nu/D = 0.01/2.6\text{x}10^{-6} = 3850$$

and

$$I_s = e^{-t/0.96}$$

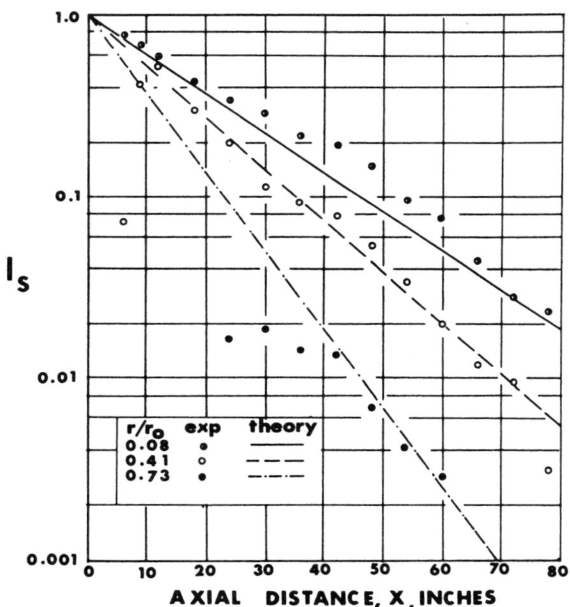

Fig. 18. Decay of I_s for selected radial positions for pipe flow mixing.

Gegner and Brodkey (1966) extended the comparison to mixing at other radial positions obtained by Lee and Brodkey (1964). The only thing that varies is u'_x that is used in Eqs. (19) and (40), and this is known, so that the time constant at any radial position can be estimated using Eq. (31). Furthermore, a normalization of all the data can be obtained by plotting I_s versus $x/\overline{U}_x\tau$. The comparisons are quite good and in the reference cited, several examples are given, the normalization plot is shown, and a tabulation of the calculated and experimental time constants are provided for twelve radial positions. Except very near the wall, the theoretical predicitions are quite close to the experimental results. This remarkably good fit between the data and Corrsin's theory is encouraging. The results show that even with the assumptions invoked, the theory gives a good estimation of mixing. This constitutes a use of isotropic theory, but not a proof. Such a good fit to the data could be misleading and several other comments are in order:

1. The data were obtained with a probe whose resolution was less than adequate and, as a result, are consistently low by as much as 30%.
2. The velocity field did not show a local isotropic range.
3. The reasonable estimates can be attributed to the required integration of the spectrum to obtain I_s. Such an integration is a smoothing process, such that, information about spectrum shape is lost.
4. The decay does not follow an exact straight line, and the least squares fit is really an average value. Nye and Brodkey (1967B) used a smaller probe, obtaining the results also shown in Fig. 18. The earlier results were consistently low. When modified on the normalized plot, this would put the predicted line below the data by a constant factor. A slight modification of one of the terms could be made to provide for the adjustment.

McKelvey *et al.* (1974) estimated λ_s directly from Eq. (24) using the Nye and Brodkey data; i.e.,

$$\lambda_s^2 = 12D/(-1/I_s)(dI_s/dt) \qquad (43)$$

This with Eq. (26) gives the time constant for mixing from the experiments. The value was 1.6 sec at 1-ft from the injection point and dropped to 1.1 sec at 9-ft. The estimated time constant is still 0.96 sec. The predicted value is close (and could be adjusted as mentioned above) but clearly should not be constant as implied from the theory. Nevertheless, the theory is adequate for a reasonable estimation of the time constant of mixing. One final comparison of interest is the macroscale, L_s. It can be experimentally determined from the spectrum from an equation parallel to Eq. (12):

$$L_s = (\pi/2a'^2)\phi_s(0) \qquad (44)$$

and can be estimated from Eq. (40) previously used for the mixing estimation. The values from the spectra decreased from 1.5 to 1.1 cm as one moved away from the injector. The estimated value was 1.0 cm. Again, the estimate is good but not constant over the entire mixing range.

One can look for an explanation for the lack of agreement in the early stage of mixing from the spectra. In the application of any theoretical approach, one would like to know if the basic assumed mechanism is indeed correct even though it may not be very important from the practical standpoint. Corrsin's theory is based upon the intuitive spectrum combination of the -5/3 and -1 regions. Figure 5 shows a -1 spectral region developed and it was quite clearly evident at the measuring positions of 6 and 9 ft from the injector, but it did not exist at 1 ft. The assumed -1 spectrum had not had the time to develop before the 6 ft and it is this same region where the estimate of the time constant is in error. A cause and effect cannot be proven, but the possibility exists.

V.B. *MIXING VESSELS*

In order to use the rough estimates for mixing vessels, one needs to know what is the characteristic dimension, L. For the mixing vessel used by Rao and Brodkey (1972A), the characteristic dimension for the jet stream was believed to be the impeller radius which was 5.1 cm.

The best experimental values were generally estimated in more than one way. For example, for the microscale, Eqs. (6), (7), and (9) had a maximum absolute deviation of 26% and a minimum of 3% with the average deviation being 13% for all positions studied. Equation (10) gave erratic results, as was to be expected, since the velocity fluctuations were not normally distributed. The actual values were about 0.1 cm near the impeller and 0.075 cm removed from the impeller. An estimate of the microscale using Eq. (18) and the velocity fluctuation data in the paper gave about 0.075 cm near the impeller and 0.1 cm away. Considering the vast difference of this system to pipe flow, the estimates can be considered excellent experimental values and estimates for other turbulence parameters (ε, L_f) can be found in the paper.

As mentioned earlier, measurements of the mixing in the impeller stream were very difficult, but two values of I_s were reported by Rao (1969). The values reported were I_s = 0.00193 at radial position of 2.5 and I_s = 0.00150 at 3.0 inches (2.0 inch impeller radius). From these two points an evaluation of the time constant of mixing could be made from Eq. (25) if one assumed a Taylor relation between the 0.5 inch seperation and the measured radial velocity (0.4 m/sec). The value obtained is 0.12 sec and, at best, can be considered a crude experimental value based on a straight line between two points. To estimate the mixing, one needs L (already suggested as 5.1 cm) and u'_x, which, over the two positions, averaged 29.3 cm/sec. With this, Eqs. (19), (31) and (40) can be used: $\tau = (1/2)[3(0.0594) + (8.8 \times 10^{-3}/2.17 \times 10^4)^{1/2} \ln 3400] = 0.092$sec. This is exactly equivalent to estimating ε from Eq. (20), k_0 from Eq. (14) and equating it to $k_{0,s}$ to get the mixing time constant. The check of 0.12 sec versus 0.092 sec is impressive. Note the ten-fold increase in the rate of mixing in the vessel as compared to the pipe centerline (inverse to the time constant).

Rao (1969) made comparisons also with the experimental mixing vessel data of Reith (1965). The Schmidt number was much lower (600) and u'_x and ε were estimated from the work of Cutter (1966). With ε known, the equivalent form to Eq. (40) involving ε and u'_x could be used. Reith's experimental values for the time constant varied from 0.018 to 0.18 sec as a function of radial position, while the estimates varied from 0.033 to 0.16 sec over the same range. In the region near the impeller, the values were about a factor of two high; even so, the check is still quite good when one considers the nature of what is being estimated.

V.C. *MULTI-JET REACTOR*

Considerable work has been done on the multi-jet reactor configuration in order to establish both the turbulence and the mixing fields. As indicated in the experimental results, the axial turbulent velocity intensity along the centerline of the reactor initially decayed as expected; however, the intensity then unexpectedly increased. The axial velocity itself decayed rapidly from the multi-jet injector head and then, after coalescence of the jets, decayed slowly over the range where reaction would occur. The nonideal flow was attributed to the formation of a large vortex or separation along the wall near the entrance of the reactor and is discussed more fully by McKelvey *et al.* (1974). Once again, correlations and spectra results were used to compute the characteristic turbulence parameters: velocity macroscale and microscale, low wave number cutoff and the kinetic energy dissipation. The results reported here are for the 100 tube injector configuration operating in water.

For the microscale, the deviation between Eqs. (6), (7), (9) and (10) was about 5% for positions close to the head and 15% several inches away. The microscale increased as the distance from the head increased, being 0.42 mm at 1.1 cm, 0.91 mm at 3.0 cm, and 1.3 mm at 8.1 cm from the head. The good check using Eq. (10) was to be expected since the velocity fluctuations were normally distributed. The macroscale from Eq. (11) ranged from 0.04 (near the head) to 0.24 mm (far from the head), which is unreasonably small when compared to 0.42 to 1.3 mm for the microscale. In contrast, the scale based on the area under the curve to the first zero crossing varied from 0.90 to 2.66 mm and the scale based on the absolute values of the first two areas (first and second zero crossings) varied from 1.76 to 5.08 mm. Both of these latter are far more reasonable. An estimate using spectra results and Eq. (12) gave for the same range: 1.85 to 9.0 mm. These compare well with the values obtained from the autocorrelation using the absolute value of the first two areas. Thus, one would conclude, in non-isotropic systems where quasi-periodic motions occur, a better definition of a macroscale would be the integration to the first zero crossing or integration of the absolute value of the correlation rather than application of Eq. (11) directly.

The kinetic energy dissipation ε, can be obtained from Eq. (13) and the values of the microscales. Thus, this is no better than the microscale estimation. Near the head, the dissipation was 5.2×10^4 cm^2/sec^3 and far from the head, 0.12×10^4 cm^2/sec^3. The high rates of mixing in the jet coalescence area are associated with high dissipation.

To attempt to calculate the parameters from the rough estimate equations [(15) to (20)] is made difficult by our total lack of knowledge of what to use for L. Near the head, L, as the radius of the injector tubes or the expanded jet radius, gives unsatisfactory results (Brodkey *et al.* 1971A). The only dimension found satisfactory by McKelvey *et al.* (1974) was the jet coalescence length. Far from the head, the only logical dimension is the reactor radius. The measured values and the estimates are compared in Table 2. The comparisons are generally good, but there are differences approaching an order of magnitude, especially far from the injector head. Actually, better results, except for L_f, would be obtained far from the head if the characteristic length would be taken as the axial distance of 8.10 cm, but there is no logical justification for this.

McKelvey *et al.* (1974) found that the scalar decay followed a -3/2 power decay law given by

$$I_s = 1.28 \times 10^{-5} \, t^{-3/2} \tag{45}$$

with time in seconds. The same power dependency had been previously observed by Gibson and Schwarz (1963) and by Keeler *et al.* (1965) and predicted by Hinze (1959). These results were obtained in approximately isotropic fields as contrasted to the multi-jet reactor field that was far from ideal. Torrest and Ranz (1970) measured the concentration decay in a system somewhat similar to the one used by Mc-Kelvey and observed a decay that fell between a (-1) and (-3/2) power-law dependency. It may be, therefore, that the decay of passive scalar fields are insensitive to the details

TABLE 2 *Multijet Reactor 100 Tube Injector Configuration*

	Equation Used		Near Head (1.11 cm)		Far from Head (8.10 cm)	
	Exp.	Est.	Exp.	Est.	Exp.	Est.
L cm			1.1		1.59	
ν cm^2/sec			0.009		0.009	
u'_x cm/sec			25.4		13.8	
λ cm	(9)	(18)	0.042	0.036	0.130	0.060
L_f cm	(11)	(16)	0.18	0.43	0.51	0.60
$\varepsilon \times 10^{-4} cm^2/sec^3$	(13)	(19)	4.7	6.5	0.11	0.72
$k_0 cm^{-1}$	(14)	(15)	1.74	1.82	0.24	1.25
τ sec	(24),(26)	(31),(40)	0.0044	0.024 0.0041*	0.067	0.064

*Expanded jet radius used for L in mixing estimated from Eq. (40). The dissipation, ε, used as above from Eq. (19).

of the velocity field. The time used in Fig. 13, which shows the scalar decay data, is the average time required for a fluid element to flow from the head to the position at which the measurement was taken; i.e., an integrated value.

One main purpose of the McKelvey *et al.* (1974) work was to demonstrate the equivalence of the mixing results in the reactor and the very fast reaction results in the same geometry as measured by Vassilatos and Toor (1965). The suggested equivalence is a result of Toor's theory discussed earlier. In order to demonstrate this, Vassilatos' data in the form of I_s, for very rapid reactions, is shown in Fig. 13 along with the mixing data without reaction. The intensity of segregation is simply calculated from the fraction conversion (F) by

$$I_s = (1 - F)^2 \tag{46}$$

The agreement is excellent for all times after the coalescence plane and since Toor's theory does not apply to coarsely non-homogeneous fields, agreement could not be expected before this position. In addition, Toor used an area averaging across the reactor for his averages; in this study, time averages were used. These two would be equal only after the coalescence plane.

Before leaving the subject of the scalar decay, the method of calculating a_0' should be mentioned. The square root of this value should be the average of the concentration fluctuations at the downstream face of the head. Since the concentrations of the jets leaving the head were not known with great accuracy, a_0' was selected as the value which yielded best agreement with Vassilatos' data. The values thus calculated were within experimental error of the area averaged values. In any case, an improper choice of a_0'

would merely add a constant to the data on a log-log plot and could not alter the shape of the decay curve.

Before turning to the rough prediction of the mixing, some comments are in order about the scalar spectra shown in Fig. 14. In spite of the large amount of background noise in the concentration signal, the one-dimensional scalar spectra were obtained. Unfortunately, much of the spectra lies beyond the cutoff frequency of the light probe and could not be obtained. However, when far enough downstream of the ejectors, a -1 power region was observed. According to Batchelor (1959) and as shown by Nye and Brodkey (1967B) and Grant *et al.* (1968), the scalar spectrum should show a -1 region for large Schmidt number experiments. Since the scalar eddies enter this reactor as large elements, it should take a finite amount of time for the velocity field to break these elements down to a small size and for the equilibrium suggested by Batchelor to develop. The time required for a scalar eddy of wave number $(\varepsilon/\nu^3)^{1/4}$ to be deformed by the straining process into one of wave number $(\varepsilon/\nu D^2)^{1/4}$ is, according to Batchelor,

$$t = -(1/\gamma)\ \ln N_{Sc} \tag{47}$$

where γ is the strain rate parameter. These times for the positions 1.10, 3.02 and 8.10 centimeters from the head are 0.0065, 0.0386 and 0.0375 seconds. The average times required for the fluid to flow from the head to these positions are 0.007, 0.032 and 0.11 seconds. Probably the only position for which the "-1" range had time to fully develop was the last one, and this position does exhibit at least one decade of the -1 range.

McKelvey *et al.* (1974) could not use the theory for decaying velocity field to estimate the scalar decay rate.

However, the analysis for a stationary velocity field as applied to specific locations was used to estimate the mixing. This is the same theory as used previously for examining scalar decay. Since the conversion of very rapid reactions is directly related to this decay, once the mixing is predicted, so is fast reaction conversion. The experimental results come directly from the defining Eq. (24) and, with Eq. (26), one can get the time constant. Equations (19), (31) and (40) give the crude estimate. Near the head, the turbulence parameters are estimated well; the actual mixing is about six times faster than predicted. Away from the head, the turbulence parameters are not estimated well, but the mixing is. The reason for this is not known, but inspection of Fig. 19, which is a plot of the mixing as

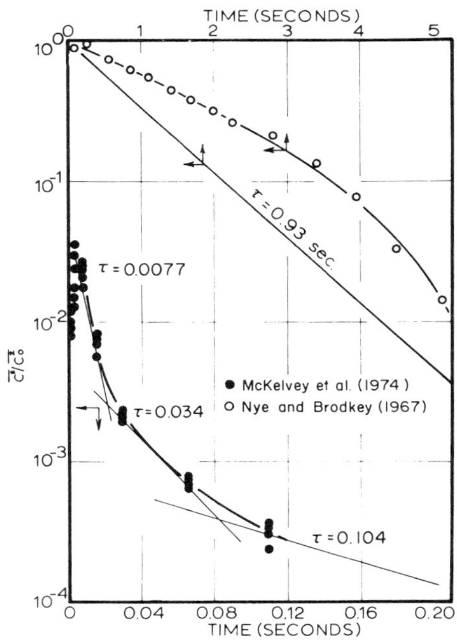

Fig. 19. Comparison of results for multijet reactor system, liquid system (Model A, ordinate is I_s, c = a).

suggested by Eq. (26), shows the extremely rapid initial mixing in the multijet reactor (97 tube model A). The near head position was selected to be right at the extremely active jet coalescence region and, quite possibly, this has much to do with not being able to predict the mixing level in this region. If one assumes the scale associated with the mixing in the region is different from that for the turbulence and is not the length of the coalescence region, but rather the expanded jet radius at the coalescence plane, then a much better estimate of the mixing is obtained. The length is used to estimate ε as was done to get the estimate in Table 2 [i.e., in Eq. (19)] and the radius is used in the estimate obtained from Eq. (40). These two are then used in Eq. (31):

$$\tau = (1/2)[3(0.00177) + (0.009/6.5 \times 10^{-4})^{1/2} \ln 3460]$$
$$= 0.0041$$

which checks well with the measured value.

One final comparison is possible and it is one that is extremely important since it involves gas mixing as measured by Ajmera (1969) by means of a fast chemical reaction between ozone and nitric oxide. For reaction close to stoichiometric conditions and for a coaxial injector pipe reactor, similar to our pipe system, Ajmera measured the fraction conversion along the center-line which can be easily translated into I_s. The pipe was a 1/8-inch I.D. tube, 4-1/2 inches long. No turbulence parameters are available, but the literature data summarized by Brodkey *et al.* (1971A) can be used. The Reynolds number is known so that the friction factor or U^*, the friction velocity, can be determined. From this, u'_x can be estimated since at the centerline

$$u'_x = 1.2\ U^* = 1.2\ \overline{U}_x \sqrt{f/2} \qquad (48)$$

Once again, using Eqs. (19), (27) and (40), the estimates can be made and compared to the experimental results; one example of which is shown in Fig. 20. Note that Eq. (27) is

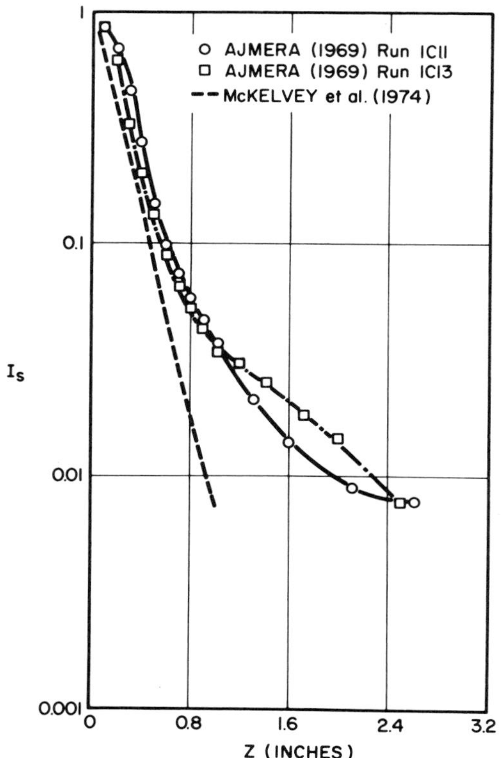

Fig. 20. Comparison of results for gas mixing and reaction.

used rather than Eq. (31) since this is a gas system. For three different Reynolds numbers, the checks were within 10%.

In Table 3 are summarized all of the mixing predictions in the various geometries considered. Apparently the analysis is reasonably valid as can be seen by the vast

TABLE 3

Comparison of τ_{exp} to τ_{est}, For All Mixing Experiments

Experiment	τ_{exp} sec	τ_{est} sec
3-inch pipe	1.1 - 1.6	0.96
stirred tank		
Rao	0.12	0.092
Reith, near impeller	0.018	0.033
Far from impeller	0.18	0.16
Multijet Liquid Reactor		
Near head	0.0044	0.067 (0.0041**)
Far from head	0.067	0.064
1/8-inch Gas Reactor*		
N_{Re} = 3480	8.6×10^{-4}	8.1×10^{-4}
6980	4.2×10^{-4}	3.8×10^{-4}
11920	2.0×10^{-4}	2.1×10^{-4}

* From fast reaction data,** see Table 2.

range of mixing times reasonably predicted (4 orders of magnitude).

VI. PERSPECTIVE

In this review, I have attempted to provide the background to what we describe as mixing. The problem, the mechanism involved, and the criteria that can be used to describe the progress of mixing have been treated in some detail. In a range of experiments, the mixing has been measured in fields where the turbulence is well established either from experiments or from available results found in the literature. Of particular importance are the parameters that describe the turbulent field for these are necessary to predict the mixing as controlled by that field. The measurement of the parameters, estimation by simple empirical means, and comparisons of the two have been provided for the geometries considered. The mixing can be measured as well as estimated from a limited knowledge of the turbulent field. Such estimations and comparisons to corresponding experiments also have been given. In the course of the review, particularly in the theory and comparison sections, the work in the field has been evaluated in terms of its contribution to our understanding of mixing and guidelines that the work can offer in real mixing operations.

Two areas have not been treated here: first, chemical reaction where both the reaction and mixing are of importance, and second, the extension by integration wherein the rates of mixing (time constant) predicted here are applied to complex flow systems. The first of these has been treated by Toor and his co-workers and by Brodkey and his and is reviewed in Toor's paper in this series. O'Brien, in Chapt. I, has covered the kinetics from the fundamental standpoint and has provided what should be improved solutions for future work. A start on the second problem has been made by Patterson, and he has reviewed this in Chapt. V.

VII. ACKNOWLEDGMENTS

The author's work reviewed herein was supported by a series of grants from the National Science Foundation. Many students worked with the author on this endeavor and their publications and these are cited in the references. The author is grateful to the other authors of this volume and to Drs. James Wallace and Stravos Nychas for their helpful suggestions. Thanks are also due to the publishers who gave permission for reproduction of the author's figures.

VIII. APPENDIX

*AN OUTLINE WITH SPECIFIC REFERENCES**

I. Types of Mixing

- A. Laminar (McKelvey, 1962; Mohr, 1959; Mohr *et al.*, 1957; Spencer and Wiley, 1957)
- B. Turbulent
  - 1. Dispersion
    - a. Basic Information (Hinze, 1959; Taylor, 1921)
    - b. Extensions (Baldwin and Mickelson, 1962; Batchelor and Townsend, 1956; Corrsin, 1959A, 1963; Saffman, 1960)
    - c. Some Applied Problems (Corrsin, 1961C; Hanratty *et al.*, 1956; Inoue, 1960; Joseph and Sender, 1962; Kofoed-Hansen, 1962)
  - 2. Turbulent Mixing (Brodkey, 1968, 1973; Danckwerts, 1953; Hughes, 1957)

II. Criteria for Mixing

- A. Scale of Segregation, L_s (Danckwerts, 1953)
- B. Intensity of Segregation, I_s (Brodkey, 1967; Danckwerts, 1953)

III. Theory

- A. Wave Number Space Analysis
  - 1. Basic Equations (Hinze, 1959; Taylor, 1938)
  - 2. Intuitive Analysis and Spectra Results (Batchelor, 1959; Batchelor *et al.*, 1959; Corrsin, 1951, 1952, 1953, 1959B; Gibson, 1968; Pao, 1965A)
  - 3. 2nd Moment Equation Analysis (Beek and Miller, 1959)
  - 4. Quasi-normal Approximation (O'Brien and Francis, 1962; Reid, 1955)

---

*The references cited herein as well as in the main text are all grouped together following this appendix outline.

5. Direct-Interaction Approximation (Lee, 1965)

B. Physical Space Analysis

1. Time Constants and Scalar Decay (Corrsin, 1957, 1964A; Rosensweig, 1964; Sutton, 1968)

2. Idealized Mixer Scale-up (Corrsin, 1957, 1964A)

3. Mixer Calculations (Brodkey, 1966A; Corrsin, 1964A; Danckwerts, 1958; Nye and Brodkey, 1967B)

4. Scale-up Problem (Corrsin, 1964A; Toor 1967)

C. Applications to Specific Equipment

1. Pipe Systems

2. Mixers (Kattan and Adler, 1972; Patterson, 1970, 1973; Rao and Edwards, 1973; Rosensweig, 1966; Trelearen and Tobgy, 1972, 1973)

3. Jets

4. Others

IV. Experimental Methods

A. Light Probe (Lee and Brodkey, 1963, 1964; Nye and Brodkey, 1967A)

B. Conductivity (Cairns and Prausnitz, 1959; Gibson and Schwarz, 1963; Keeler *et al.*, 1965; Lamb *et al.*, 1960; Torrest and Ranz, 1969)

C. Light Scattering (Becker *et al.*, 1967B; Hawthorne *et al.*, 1949; Rosensweig *et al.*, 1961; Schwartz, 1963)

D. Kinetics (Danckwerts, 1953, 1957, 1958)

V. Experimental Results and Comparison to Theory

A. Pipe Flow (Becker *et al.*, 1966, 1967B; Brodkey, 1966A, 1966B; Brodkey *et al.*, 1971A; Danckwerts, 1958; Gegner, 1965; Laufer, 1954; Lee, 1962; Lee and Brodkey, 1964; Nye and Brodkey, 1967B)

B. Mixers (Cutter, 1966; Fort; Rao, 1969; Rao and Brodkey, 1972A, 1972B; Reith, 1965)

C. Jets (Becker *et al.*, 1967A; McKelvey, 1968; McKelvey *et al.*, 1974; Rosensweig *et al.*, 1961)

D. Others (Ajmera, 1969; Asjϕrnsen and Klovning, 1968; Grant *et al.*, 1968; Mao and Toor, 1971; McKelvey *et al.*, 1974; Toor and Singh, 1973; Vassilatos and Toor, 1965; Yieh, 1970; Zakanycz, 1971)

VI. Extensions to Kinetics

A. Initially Separated Reactants

1. Fast Reactions (Dopazo and O'Brien, 1973; Gibson and Libby, 1972; Keeler *et al.*, 1965; Libby, 1972; O'Brien, 1969, 1971A, 1971B; O'Brien and Lin, 1972; Toor, 1962, 1969; Toor and Singh, 1973)

2. In Between Reactions (Harris and Srivastava, 1968; Kattan and Adler, 1967; Mao and Toor, 1970, 1971; Rao and Dunn, 1970; Rao and Edwards, 1971; Vassilatos and Toor, 1965)

3. Slow Reactions

B. Initially Together Reactants (Corrsin, 1953, 1958, 1961A, 1961B, 1964B; Pao, 1964, 1965B)

C. Self Mixing (Bischoff and McCracken, 1966; Bischoff, 1966; Danckwerts, 1958; Levenspiel, 1962; Levenspiel and Bischoff, 1963; Zwietering, 1959)

IX. REFERENCES

Ajmera, P.V. (1969) Ph.D. Thesis, Carnegie-Mellon University.

Asbjǿrnsen, O.A. and Klovning, M. (1968) *Chem Engr. Sci. 23,* 1053.

Baldwin, L.V. and Mickelsen, W.R. (1962) *J. Eng. Mech. 88,37,* 151.

Batchelor, G.K. (1959) *J. Fluid Mech. 5,* 113.

Batchelor, G.K., Howells, I.D. and Townsend, A.A. (1959) *J. Fluid Mech. 5,* 134.

Batchelor, G.K. and Townsend, A.A. (1956) In SURVEYS IN MECHANICS (G.K. Batchelor and R.M. Daries, eds.) p. 352. Cambridge University Press, London and New York.

Becker, H.A., Rosensweig, R.E. and Gwozdz, J.R. (1966) *A.I.Ch.E.J. 12,* 964.

Becker, H.A., Hottel, H.C. and Williams, G.C. (1967A) *11th Symp. on Combustion* 791.

Becker, H.A., Hottel, H.C. and Williams, G.C. (1967B) *J. Fluid Mech. 30,* 259, 285.

Beek, J., Jr. and Miller, R.S. (1959) *Chem. Eng. Progr. Symposium Ser. No. 25, 55,* 23; private communication.

Bischoff, K.B. (1966) *Ind. Eng. Chem. 58 #11,* 18.

Bischoff, K.B. and McCracken, E.A. (1966) *Ind. Eng. Chem. 58 #7,* 18.

Brodkey, R.S. (1966A) *A.I.Ch.E.J. 12,* 403.

Brodkey, R.S. (1966B) Fluid Motion and Mixing, Chap. 2 in Vol. 1 of MIXING: THEORY AND PRACTICE, Uhl and Gray eds., Academic Press Inc., New York.

Brodkey, R.S. (1967) THE PHENOMENA OF FLUID MOTIONS, Addison-Wesley Publishing Company, Inc., Reading, Mass.

Brodkey, R.S. (1968) Prof. Dev. Lect. *1* #2, W. Va. Univ., Nitro, West Va.

Brodkey, R.S., Cohen, M.F., Capt. Knox, J.S., McKee, G.L., McKelvey, K.N., Rao, M.A., Yieh, H.N., and Zakanycz, S., (1971A) PROC. SYMP. TURBULENCE MEASUREMENTS IN LIQUIDS, Univ. of Missouri-Rolla.

Brodkey, R.S., Hershey, H.C., and Corino, E.R. (1971B) PROC. SYMP. TURBULENCE MEASUREMENTS IN LIQUIDS Univ. of Missouri-Rolla.

Brodkey, R.S. (1973) in FLUID MECHANICS OF MIXING, ASME 1.

Cairns, E.J. and Prausnitz, J.M. (1959) *Ind. Eng. Chem. 51,* 1441; (1960)*A.I.Ch.E.J. 6,* 400; (1960) ibid 554.

Corino, E.R. and Brodkey, R.S. (1969) *J. Fluid Mech. 37,* 1.

Corrsin, S. (1951) *J. Appl. Phys. 22,* 469.

Corrsin, S. (1952) *J. Appl. Phys. 23,* 113.

Corrsin, S. (1953) In PROCEEDINGS OF THE FIRST IOWA THERMODYNAMICS SYMPOSIUM, Iowa State University, Ames.

Corrsin, S. (1957) *A.I.Ch.E.J. 3,* 329.

Corrsin, S. (1958) *Phys. Fluids 1,* 42.

Corrsin, S. (1959A) *Advances in Geophys. 6,* 161, 441.

Corrsin, S. (1959B) *J. Aeronaut. Sci. 18,* 417.

Corrsin, S. (1961A) In PROCEEDINGS SYMPOSIUM ON FLUID DYNAMICS AND APPLIED MATHEMATICS pp. 105-124. University of Maryland. Gordon and Breach, New York.

Corrsin, S. (1961B) *J. Fluid Mech. 11,* 407.

Corrsin, S. (1961C) *Am. Scientist 49,* 300.

Corrsin, S. (1963) *Atmospheric Sci. 20,* 115.

Corrsin, S. (1964A) *A.I.Ch.E.J. 10,* 870.

Corrsin, S. (1964B) *Phys. Fluids 7,* 1156.

Cutter, L.A. (1966) *A.I.Ch.E.J. 12,* 35.

Danckwerts, P.V. (1953) *Appl. Sci. Research A3,* 279.

Danckwerts, P.V. (1957) *Chem. Eng. Sci. 7,* 116.

Danckwerts, P.V. (1958) *Chem. Eng. Sci. 8,* 93.

Dopazo, C. and O'Brien, E.E. (1973) in FLUID MECHANICS OF MIXING, ASME 117; *Phys. Fluids 16,* 2075.

Fort, I. *Collection of Czech. Chem. Comm.,* a series of 36 articles covering all aspects of turbulent mixing vessels.

Gegner, J.P. (1965) M.S. Thesis, The Ohio State Univ., Columbus, Ohio.

Gegner, J.P. and Brodkey, R.S. (1966) *A.I.Ch.E.J. 12,* 817.

Gibson, C.H. and Schwarz, W.H. (1963) *J. Fluid Mech. 16,* 357, 365.

Gibson, C.H. (1968) *Phys. Fluids 11,* 2305.

Gibson, C.H. and Libby, P.A. (1972) *Comb. Sci. and Tech. 6,* 29.

Grant, H.L., Hughes, B.A., Vogel, W.M. and Moilliet, A. (1968) *J. Fluid Mech. 34,* 423.

Hanratty, T.J., Latimen, G. and Wilhelm, R.H. (1956) *A.I.Ch. E.J. 2,* 372.

Harris, I.J. and Srivastava, R.D. (1968) *Can.J. of Chem. Engr. 46,* 66.

Hawthorne, W.R., Weddell, D.S. and Hottel, H.C. (1949) THIRD SYMPOSIUM ON COMBUSTION, FLAME AND EXPLOSION p. 266. The Williams and Wilkins Company, Baltimore.

Hinze, J.O. (1959) TURBULENCE, McGraw-Hill, Inc., New York, 1959.

Hughes, R.R. (1957) *Ind. Eng. Chem. 49,* 947.

Inoue, Eiichi (1960) PROC. 10TH JAPAN NATL. CONGR. APPL. MECH. p. 217; Metol. Res. Notes 11, 332.

Joseph, J. and Sender, H. (1962) *J. Geophys. Res. 67,* 3217.

Kattan, A. and Adler, R.J. (1967) *A.I.Ch.E.J. 13,* 580.

Kattan, A. and Adler, R.J. (1972) *Chem. Engr. Sci. 27,* 1013.

Keeler, R.N., Petersen, L.E. and Prausnitz, J.M. (1965) *A.I. Ch.E.J. 11,* 221.

Kofoed-Hansen, O. (1962) *J. Geophys. Research 67,* 3217.

Lamb, D.E., Manning, F.S. and Wilhelm, R.H. (1960) *A.I.Ch. E.J. 6,* 682.

Laufer, J. (1954) N.A.C.A. REPORT, 1174.

Lee, J. (1962) Ph.D. Thesis, The Ohio State Univ., Columbus, Ohio.

Lee, Jon and Brodkey, R.S. (1963) *Rev. Sci. Instr. 34,* 1086.

Lee, Jon and Brodkey, R.S. (1964) *A.I.Ch.E.J. 10,* 187.

Lee, Jon (1965) *Phys. Fluids 8,* 1647.

Levenspiel, O. (1962) CHEMICAL REACTION ENGINEERING, Wiley, New York.

Levenspiel, O. and Bischoff, K.B. (1963) ADVANCES IN CHEMICAL ENGINEERING *4,* 95.

Libby, P.A. (1972) *Comb. Sci. and Tech. 6,* 23.

Mao, K.W. and Toor, H.L. (1970) *A.I.Ch.E.J. 16,* 49.

Mao, K.W. and Toor, H.L. (1971) *Ind. Eng. Chem. Fund. 10,* 192.

McKelvey, J.M. (1962) POLYMER PROCESSING, John Wiley & Sons, Inc., New York.

McKelvey, K.N., Yieh, H.N., Zakanycz, S. and Brodkey, R.S. (1974) (submitted to *A.I.Ch.E.J.*).

McKelvey, K.N. (1968) Ph.D. Thesis, The Ohio State Univ., Columbus, Ohio.

Mohr, W.D. (1959) In PROCESSING OF THERMOPLASTIC MATERIALS (Bernhardt, ed.), p. 117, Reinhold, New York.

Mohr, W.D., Saxton, R.L. and Jepson, C.H. (1957) *Ind. Eng. Chem. 49,* 1855; ibid (1957) 1857.

Nychas, S.G., Hershey, H.C. and Brodkey, R.S. (1973) *J. Fluid Mech. 61,* 513.

Nye, J.O. and Brodkey, R.S. (1967A) *Rev. Sci. Insts. 38,* 26.

Nye, J.O. and Brodkey, R.S. (1967B) *J. Fluid Mech. 29,* 151.

O'Brien, E.E. (1969) *Phys. Fluids 12,* 1999.

O'Brien, E.E. (1971A) *Phys. Fluids 14,* 1326.

O'Brien, E.E. (1971B) *Phys. Fluids 14,* 1804.

O'Brien, E.E. and Francis, G.C. (1962) *J. Fluid Mech. 13,* 369.

O'Brien, E.E. and Lin, C.H. (1972) *Phys. Fluids 15,* 931.

Pao, Y.H. (1964) *A.I.A.A. Journal 2,* 1550.

Pao, Y.H. (1965A) *Phys. Fluids 8,* 1063.

Pao, Y.H. (1965B) *Chem. Engr. Sci. 20,* 665.

Patterson, G.K. (1970) PROC. CHEMECA 70, Melborne, Australia.

Patterson, G.K. (1973) In FLUID MECHANICS OF MIXING, ASME 31.

Rao, D.P. and Dunn, I.J. (1970) *Chem. Engr. Sci. 25,* 1275.

Rao, D.P. and Edwards, L.L. (1971) *A.I.Ch.E.J. 17,* 1264.

Rao, D.P. and Edwards, L.L. (1973) *Chem. Engr. Sci. 28,* 1179.

Rao, M.A. (1969) Ph.D. Thesis, The Ohio State Univ., Columbus, Ohio.

Rao, M.A. and Brodkey, R.S. (1972A) *Chem. Engr. Sci. 27,* 137.

Rao, M.A. and Brodkey, R.S. (1972B) *Chem. Engr. Sci. 27,* 2199.

Reid, W.H. (1955) *Proc. Cambridge Phil. Soc. 51,* 350.

Reith, Ir.T. (1965) *A.I.Ch.E.-I.Ch.E. Symp. Series 10,* 14.

Rosensweig, R.E. (1964) *A.I.Ch.E.J. 10,* 91.

Rosensweig, R.E. (1966) *Can. J. Chem. Engr. 44,* 255.

Rosensweig, R.E., Hottel, H.C. and Williams, G.C. (1961) *Chem. Eng. Sci. 15,* 111.

Saffman, P.G. (1960) *J. Fluid Mech. 8,* 273.

Spencer, R.S. and Wiley, R.M. (1957) *J. Colloid Sci. 6,* 133.

Schwartz, L.M. (1963) *Chem. Eng. Sci. 18,* 223.

Sutton, G.W. (1968) *Phys. Fluids 11,* 671.

Taylor, G.I. (1921) *Proc. London Math. Soc. 20,* 196.

Taylor, G.I. (1938) *Proc. Roy. Soc. (London) 164A,* 15, 476.

Toor, H.L. (1962) *A.I.Ch.E.J. 8,* 70.

Toor, H.L. (1967) *A.I.Ch.E.J. 13,* 616.

Toor, H.L. (1969) *Ind. Eng. Chem. Fund. 8,* 655.

Toor, H.L. and Singh, M. (1973) *Ind. Eng. Chem. Fund 12,* 448.

Torrest, R.S. and Ranz, W.E. (1969) *Ind. Eng. Chem. Fund. 8,* 810.

Torrest, R.S. and Ranz, W.E. (1970) *A.I.Ch.E.J. 16,* 930.

Trelearen, C.R. and Tobgy, A.H. (1972) *Chem. Engr. Sci. 27,* 1653.

Trelearen, C.R. and Tobgy, A.H. (1973) *Chem. Engr. Sci. 28,* 413.

Vassilatos, G. and Toor, H.L. (1965) *A.I.Ch.E.J. 11,* 666.

Wallace, J.M., Eckelmann, H. and Brodkey, R.S. (1972) *J. Fluid Mech. 54,* 39.

Yieh, H-N. (1970) Ph.D. Thesis, The Ohio State Univ., Columbus, Ohio.

Zakanycz, S. (1971) Ph.D. Thesis, The Ohio State Univ, Columbus, Ohio.

Zwietering, Th.N. (1959) *Chem. Eng. Sci. 11,* 1.

The Non-Premixed Reaction: $A + B \longrightarrow$ Products

H. L. Toor

Chapter III

The Non-premixed Reaction: A + B → Products

H. L. TOOR

Carnegie-Mellon University
Pittsburgh, Pennsylvania 15213

I. INTRODUCTION

This review is concerned with predicting the time average rate of a chemical reaction (or the conversion) in a turbulent fluid in which the concentrations of the reacting species are fluctuating with time. The problem arises when chemical species which react with each other are mixed turbulently, but is interesting only when the time scale of the reaction is comparable to, or less than, the time scale of the mixing process (for otherwise the fluctuations play no role).

When a chemical reaction which is homogeneous in the chemical sense takes place in a turbulent fluid, the local instantaneous rate of reaction is described by the normal laws of homogeneous chemical kinetics. However, the local time average rate of reaction, $\overline{R}_1$, the *mean rate,* may be greater than, equal to, or less than the homogeneous rate at the local time average concentration, the *homogeneous mean rate*. Although the above statements are true whether or not the system is isothermal, in this review, only (effectively) isothermal situations are considered, i.e., the reaction velocity constant does not fluctuate.

For the purposes of this discussion, a homogeneous solution or mixture is one in which the r.m.s. concentration fluctuations of all species are zero. The mean rate obviously equals the homogeneous mean rate in homogeneous solutions and this is true even in nonhomogeneous solutions if the reaction is first order since

$$A \rightarrow \text{Products}$$

$$R_A = -kA \tag{1}$$

$$\overline{R}_A = -k\overline{A}$$

where A is the mass or concentration fraction or the concentration since a constant density system is being considered.

In nonhomogeneous solutions, the mean rate of a type I second-order reaction of the form A + A → product is greater than the homogeneous mean rate, while the mean rate of a type II second-order reaction of the form A + B → product may be greater or less than the homogeneous mean rate, depending upon how the reactants are introduced; greater if the reactants are premixed, less if they are not. The above assertions are easily justified by Eqs. (2) and (3),

$$A \rightarrow \text{Products}$$

$$R_A = -kA^2 \tag{2}$$

$$\overline{R}_A = -k(\overline{A}^2 + a'^2)$$

$$A + nB \rightarrow \text{Products}$$

$$R_A = -kAB \tag{3}$$

$$\overline{R}_A = -k(\overline{A}\,\overline{B} + \overline{ab})$$

noting that premixing gives a positive correlation between the fluctuating reactant concentrations while lack of premixing gives a negative correlation. The r.m.s. fluctuations are denoted by $a' = \sqrt{\overline{a^2}}$.

The deviation from the homogeneous mean rate depends upon the rate of the reaction relative to the rate of mixing. Three divisions are convenient: very rapid reactions, rapid reactions, and slow reactions. They correspond, respectively, to reaction rates much faster than the rate of mixing, reaction rates of the same order as the rate of mixing, and reaction rates much slower than the rate of mixing. The latter case is the simplest, and most reactor design studies have been concerned with this problem. It is under reasonably good control. It is noted that in the flow methods of studying *rapid* chemical reactions in solution, the reactions are slow in the above sense since the experiments are designed to eliminate inhomogeneities in concentration.

Very rapid and rapid reactions present a different problem, for here the detailed turbulent motion can have a profound effect on the rate of reaction, especially with type II second-order reactions. Here, two molecular species must be brought into intimate contact by molecular diffusion, aided and abetted by the turbulent motion.

A premixed reactor is now an impossibility; the mixer is the reactor and the problem is to determine the mean rate of reaction, hence the mean conversion, under these conditions.

Although these problems are interesting because they represent a new class of problems with large, experimentally accessible effects, which also challenge turbulence theory, their industrial importance is not clear. This is partly due to the black art approach which appears to be used industrially in handling mixing problems of this type (partly because of the lack of data on mixing). Probably more important, however, is the fact that mixing becomes more important as reaction speed increases and for high speed reactions it is not costly to overdesign in order to avoid the mixing problem.

Perhaps the more important problem is the effect of mixing on product distribution when there is more than one reaction, but it has seemed reasonable to attempt to understand the single reactions discussed here before attempting the more difficult multiple reactions.

The clearly important industrial reaction-mixing problem is combustion, which, of course, is hardly isothermal, but, again, the understanding of the simple isothermal situation seems to be a prerequisite to understanding this more difficult problem.

At the heart of the problem of predicting $\overline{R}_A$ is the predicition of $\overline{ab}$, a term clearly related to the turbulent mixing. The prediction of this term is a proper subject for the statistical theory of turbulence, but in the present absence of a general solution, it is useful to consider limiting solutions which can be readily obtained.

II. THE GENERAL PROBLEM

It is reasonable to assume incompressible flow and Fickian diffusion for almost all situations. Hence, the instantaneous equations of change are

$$(\partial A/\partial t) + \vec{U}\cdot\vec{\nabla}A = D_A\nabla^2 A + R_A \tag{4}$$

$$(\partial B/\partial t) + \vec{U}\cdot\vec{\nabla}B = D_B\nabla^2 B + nR_A \tag{5}$$

the time averaged forms are

$$(\partial\overline{A}/\partial t) + \vec{U}\cdot\vec{\nabla}\overline{A} = \vec{\nabla}\cdot D_{AT}\vec{\nabla}\overline{A} + \overline{R}_A \tag{6}$$

$$(\partial \overline{B}/\partial t) + \vec{U}\cdot\vec{\nabla}\overline{B} = \vec{\nabla}\cdot D_{BT}\vec{\nabla}\overline{B} + n\overline{R}_A \tag{7}$$

It is these later equations one has to work with eventually, if only in the simplified form of the Hougen and Watson equation, but they are intractable **because of the** unknown term $\overline{ab}$ which appears in $\overline{R}_A$, Eq. (3). Hence, we return to Eqs. (4) and (5) and note that if $D_A = D_B$, then the Burke and Schumann (1928) transformation; multiplication of Eq. (4) by n and subtraction of the resulting equation from Eq. (5) gives

$$(\partial J/\partial t) + \vec{U}\cdot\vec{\nabla}J = D\nabla^2 J \tag{8}$$

where

$$J = nA - B \tag{8a}$$

so that the quantity J is conserved instantaneously during the reaction at every point in the system. The assumption $D_A = D_B$ will be examined later in the light of experiment. It is a valid approximation. Thus, if the boundary conditions on Eq. (8) are of the proper form, it is possible to relate J to an instantaneous concentration in a pure turbulent (or non-turbulent) mixing problem, and Toor (1962) gives a number of examples. Jumping ahead a bit, we can note that if J (or some suitable statistical value of J) can be determined in this way, we will still require one further piece of information to extract A and B from the linear combination of A and B represented by J.

As an example, we consider the type of reactor we have been studying in which species A and B each enter a tubular reactor through one or more jets. If we label the jets through which species A enter as 1 and those through which species B enter as 2, then the boundary conditions on Eq. (8) are

at the 1 inlets

$$A = A_0, \quad B = 0, \quad J = J_1 = nA_0$$

at the 2 inlets (9)

$$A = 0, \quad B = B_0, \quad J = J_2 = -B_0$$

and on all other boundaries

$$\partial A/\partial y = \partial B/\partial y = \partial J/\partial y = 0$$

Let

$$(J - J_2)/(J_1 - J_2) = (J + B_0)/(nA_0 + B_0) = Y \quad (10)$$

Y is the dimensionless concentration and can be a function of time and position. Then,

at the 1 inlets

$$Y = 1$$

(11)

at the 2 inlets

$$Y = 0$$

and with Eq. (8)

$$(\partial Y/\partial t) + \vec{U}\cdot\vec{\nabla}Y = D\nabla^2 Y \quad (12)$$

Now suppose species B is not present in the inlet streams to the 2 jets (the flow rates are unchanged since reactants are assumed dilute). Then species A does not react, rather A acts as a tracer in the mixing of the 1 and 2 streams. If we let

$$Y_m = A_m/A_0 \quad (13)$$

where the subscript - m denotes a tracer. In Chapt. II, the subscript was not needed since only mixing was considered there. It is seen that Y_m satisfies precisely the same conditions as Y [Eq. (12)] so

$$(J - J_2)/(J_1 - J_2) = (J + B_0)/(nA_0 + B_0) = \tag{14}$$
$$(nA - B + B_0)/(nA_0 + B_0) = Y_m$$

Thus

$$(\overline{J} - J_2)/(J_1 - J_2) = (n\overline{A} - \overline{B} + B_0)/(nA_0 + B_0) = \overline{Y}_m \tag{15}$$

and subtracting Eq. (15) from Eq. (14) gives

$$j/(J_1 - J_2) = (na - b)/(nA_0 + B_0) = Y_m \tag{16}$$

where $J = \overline{J} + j$ and $Y_m = \overline{Y}_m + y_m$ as normally used. Hence

$$j'/(J_1 - J_2) = y_m' \tag{17}$$

where the prime denotes r.m.s. value, and

$$j'/j_0' = y_m'/y_{m,0}' = d(Z, N_{Re}, w_A/w_B, N_{Sc}) \tag{18}$$

The intensity of segregation, $\sqrt{I_s}$, is the same as d, Z is a dimensionless axial distance and w is a mass flow rate. Brodkey discussed methods of predicting d or I_s in Chapt. II. There, the subscript -m is not used since only mixing without reaction is considered. Also from Eqs. (10) and (11)

$$(\overline{J} - J_2)/(J_1 - J_2) = \overline{Y} = \overline{Y}_m \tag{19}$$

These equations relate J and various statistics of J to the corresponding behavior of a tracer in a non-reacting system which itself is a measure of the mixing in the non-reacting system. Since our goal is a determination of $\overline{A}$ and $\overline{B}$, further information is clearly needed to resolve these quantities. We now turn to the two limits where this resolution is possible without a statistical theory of turbulence.

III. THE VERY RAPID REACTION LIMIT

In any turbulent mixing process, the reaction between species A and B must be preceeded by a close contact between the molecules of A and B and on this fine scale (and moving with the fluid) there must be some reaction speed (reaction velocity constant) above which species A and B cannot co-exist for some sufficiently small time. (More precisely, the volume of coexistence can be made as small as desired.) Reaction surfaces must develop which separate regions of fluid which contain species A from regions which contain species B and the reaction rate at the reaction surfaces is controlled by molecular diffusion to the reaction surfaces. Thus at a fixed point in the fluid, we would see no reaction ($R_A = 0$) except when reaction surfaces sweep by. This is the same condition described in Chapt. I, Fig. 2. The time mean of these infinitely large pulses gives the finite $\overline{R}_A$ in Eqs. (4) and (5) and from Eq. (3) we see that at this limit, where $k \to \infty$,

$$\overline{ab}/\,\overline{A}\,\overline{B} = -1 \qquad (20)$$

The resolution of $\overline{A}$ and $\overline{B}$, which is equivalent to determining the limiting value of $\overline{R}_A$, is obtained (Toor, 1962) by a method first used by Hawthorne, Weldel and Hottel (1949). It includes the assumption that in turbulent mixing, reaction concentration fluctuations are distributed around the mean in a Gausian manner, but this assumption can be relaxed (O'Brien, 1971).

Since both reaction species cannot be present at the same point at the same time, the instantaneous concentration at a point must be either A or B. Thus, from Eq. (8a), $J > 0$ whenever species A is present and when $J < 0$, $A = 0$. Furthermore, when species A is present, $J = nA$.

Hence,

$$\begin{aligned} &\text{when } J < 0,\ A = 0 \\ &\text{when } J > 0,\ A = J/n \end{aligned} \tag{21}$$

$\overline{J}$ at the point under consideration can be positive or negative. Consider the former case. (The same final result is obtained in either case.) Then from the above

$$\begin{aligned} &\text{when } j < -\overline{J},\ A = 0 \\ &\text{when } j > -\overline{J},\ A = J/n \end{aligned} \tag{22}$$

Thus

$$\overline{A} = (1/n)\int_0^\infty J\ \phi(J)\ dJ = (1/n)\int_{-\overline{J}}^\infty (J + j)\ \phi(j)\ dj \tag{23}$$

and since j and a_m are linearly related from Eq. (16), the assumed Gaussian distribution of a_m implies that $\phi(j)$ is a Gaussian distribution. O'Brien (1971) has considered the situation in which the distribution is not Gaussian.

The result of integrating Eq. (23) with this distribution is (Toor, 1962)

$$n\overline{A}/\ \overline{J} = 1 + (\gamma_j/\sqrt{2})\ \text{ierfc}(1/\gamma_j\sqrt{2}) \tag{24}$$

where $\gamma_j = j'/\overline{J}$ and $\overline{B}$ is given by the time average form of Eq. (8a)

$$\overline{B} = n\overline{A} - \overline{J} \tag{25}$$

The scheme for computing $\overline{A}$ and $\overline{B}$ is as follows:

1. Determine the time average concentration field of a tracer in the absence of reaction by analysis [solve the time averaged form of Eq. (12) using boundary conditions {Eq. (11)}] or by experiment.
2. Obtain $\overline{J}$ from Eq. (15).
3. Obtain y_m' from a tracer experiment or theory.

4. Obtain j' from Eq. (17).
5. Compute $\overline{A}$ and $\overline{B}$ from Eqs. (24) and (25).

The most extensive confirmation of the above results has been obtained in the one-dimensional case.

IV. ONE-DIMENSIONAL REACTOR

In the one-dimensional reactor (idealized), say a tubular reactor, mean quantities vary only in the axial direction (uniformity on a coarse scale). Then $\overline{J} = \overline{J}_0$ since $\overline{J}$ is conserved, and the set of equations listed above can be reduced to (Toor, 1962)

$$\overline{A}/\overline{A}_0 \equiv X = (\beta - 1)[1 + (\gamma_{j0} d/\sqrt{2})\ \mathrm{ierfc}(1/\gamma_{j0} d\sqrt{2})] \tag{26}$$

Since $X = d = \sqrt{I_s} = 1$ at the reactor inlet, Eq. (26) gives γ_{j0}

$$(\gamma_{j0}/\sqrt{2})\ \mathrm{ierfc}(1/\gamma_{j0}\sqrt{2}) = \beta/(\beta - 1) \tag{27}$$

Also from Eq. (25)

$$\overline{B} - \overline{B}_0 = n(\overline{A} - \overline{A}_0) \tag{28}$$

When the reactants are fed in stoichiometric proportions, $\beta = 1$ and Eqs. (27) and (28) take on the simple and useful form

$$\overline{A}/\overline{A}_0 \equiv X = d = \sqrt{I_s} \tag{29}$$

Thus, in the one-dimensional reactor with a stoichiometric feed, the fraction unreacted equals the fraction unmixed in the absence of reaction. It should be noted that d is the same whether the tracer used to measure d is introduced into the 1 jets or 2 jets (Toor, 1969).

Alternate forms in terms of fractional conversion and accomplished mixing are

$$F = 1 - X = 1 - d = 1 - \sqrt{I_s} = \eta \tag{29a}$$

Eqs. (26) and (29) can be tested experimentally and the first test was carried out by Keeler, Petersen and Prausnitz (1965). They carried out experiments in which 156 or 37 hypodermic needles were used to inject either an electrolyte into a mainstream of water or acetic acid into a mainstream of ammonium hydroxide (a very rapid reaction). Wire mesh grids of various diameters were introduced downstream of the hypodermic needles to produce a known turbulence. In the mixing experiments, d was measured as a function of distance along the reactor with a microprobe and in reaction experiments under identical hydrodynamic conditions, X was also measured as a function of distance by measuring the mean conductivity.

Figures 1,2 and 3 show the results of these authors.

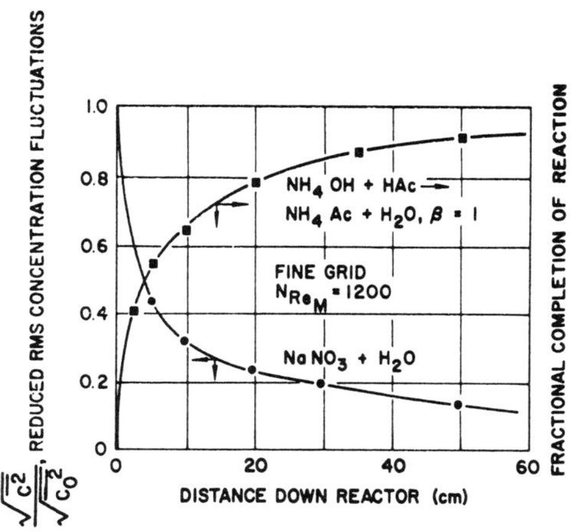

Fig. 1. Rate of approach to completion of reaction with decay of concentration fluctuations (ordinate is d or $\sqrt{I_s}$). From Keeler, Petersen and Prausnitz (1965).

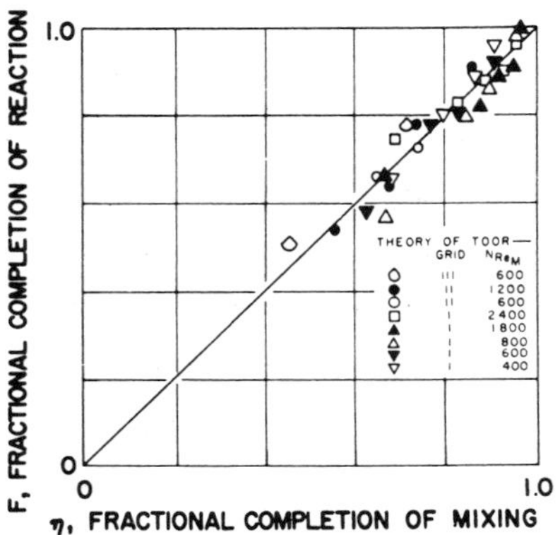

Fig. 2. Theoretical predictions and experimental results, β = 1. From Keeler, Petersen and Prausnitz (1965).

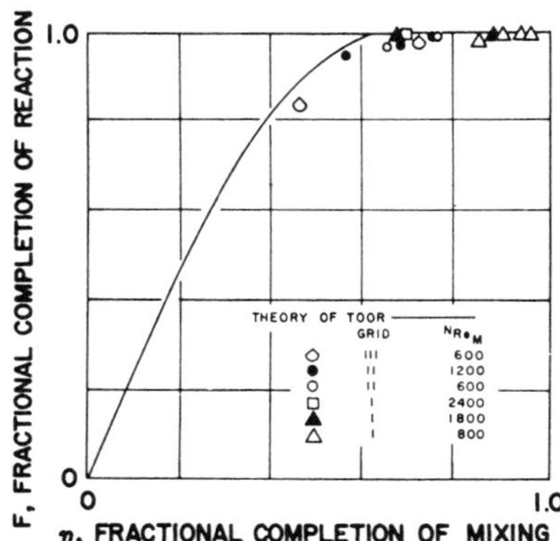

Fig. 3. Theoretical predictions and experimental results, β = 10. From Keeler, Petersen and Prausnitz (1965).

It should be noted that Eqs. (26) to (29) are valid for any starting point along the reactor. In these experiments, the starting point was taken downstream of the grid after uniformity on a coarse scale was obtained and estimates were made of the initial values of d and $\overline{A}$ at these points.

Thus there is some uncertainty in these results because of these estimates, but they do offer reasonably good support of the analysis.

Vassilatos and Toor (1965) used a mixing device (Fig. 4 and in more detail in Chapt. II) which is much more

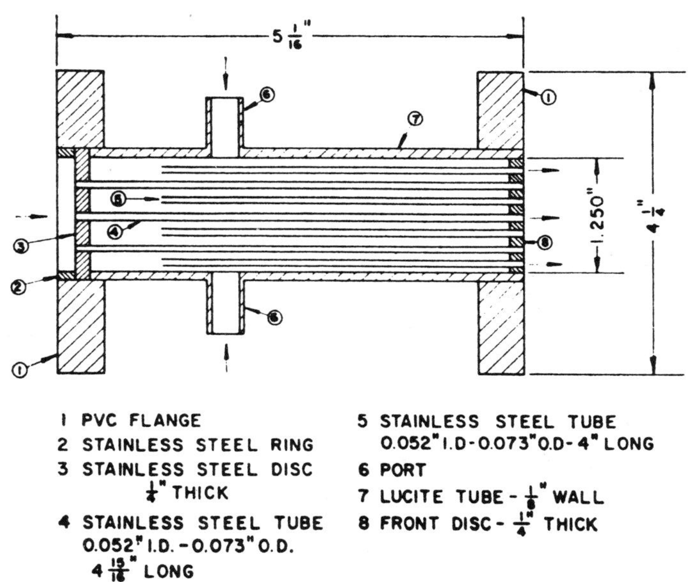

Fig. 4. Mixing device. Vassilatos and Toor (1965).

isotropic than that used by Keeler, *et al.* They did not measure d directly, but used Eq. (29) to determine d for a number of very rapid stoichiometric ionic reactions in which k was the order of 10^{11} ℓ/mole sec. and n was 1 or 2. These systems are listed in Table 1. They measured the conversion by following the temperature rise along the reactor (approximately 0.2°C total temperature change),

$$\Delta\overline{A} = (\rho C_p/\Delta H)\Delta\overline{T} \tag{30}$$

TABLE 1 *Reaction Systems Used by Vassilatos and Toor*

Reactants	Reaction Velocity Constant k, ℓ/mole sec., 30°C
HCl-NaOH	
HCl-LiOH	$\approx 10^{11}$
HOOCCOOH-2LiOH	
HCOOH-LiOH	
CO_2-2NaOH	1.24×10^4
CO_2-nNH_3 (n varies from 1 to 2)	5.85×10^2
$HCOOCH_3$-NaOH	4.70×10

All the systems gave identical results in accord with the prediction that X is independent of inlet concentration and n at a fixed β. Since the diffusivities of the reactants varied somewhat from system to system, these results tend to confirm the assumption that the diffusivities of the reactants can be taken as equal and that d is not a sensitive function of molecular diffusivity.

Note that the mean rates of these reactions are the order of 10^{-7} times the rate which would be obtained if the reactions were homogeneous, so from Eq. (3),

$$\overline{ab}/\ \overline{A}\ \overline{B} = 1 - \text{order}\ (10^{-7}) \tag{31}$$

The reactions are clearly at the diffusion controlled limit.

The measured values of d were then used with Eqs. (26) and (27) to predict X for different values of β and Fig. 5 shows the good agreement between predicted and expected values [even though McKelvey, *et al.*(1974) showed that there is a rather peculiar velocity field with this

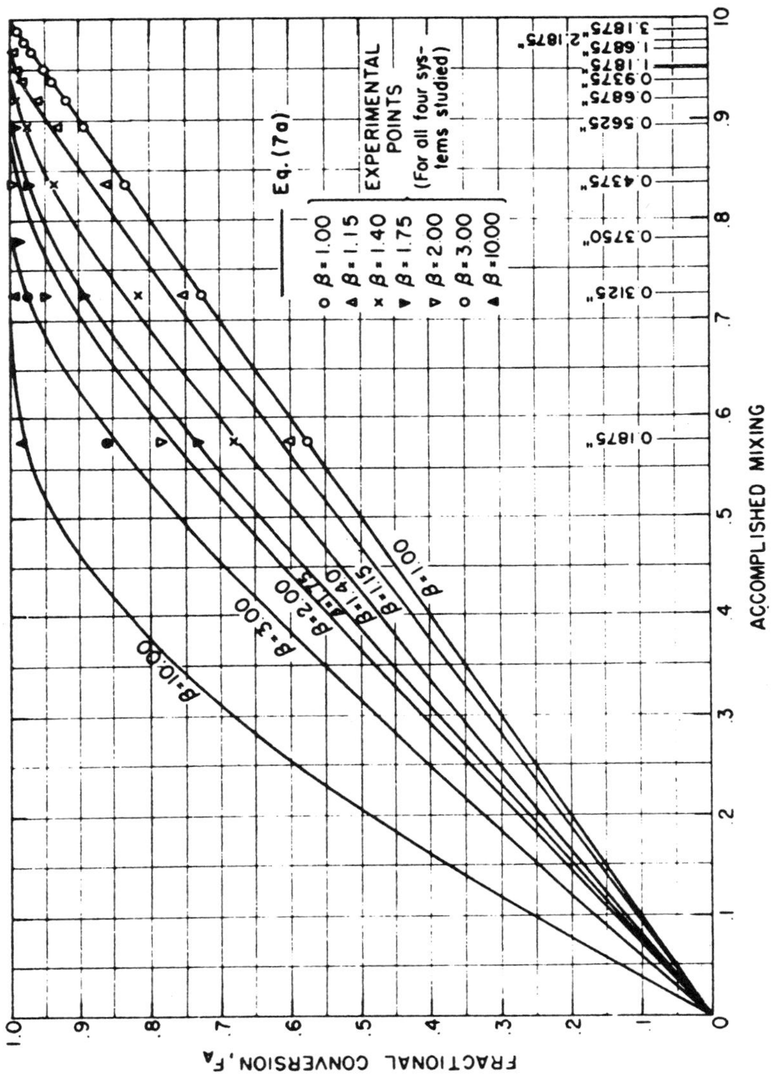

Fig. 5. Fractional conversion ($F_A = F$) vs. accomplished mixing ($1 - d = 1 - \sqrt{I_s}$), very rapid reactions. From Vassilatos and Toor (1965).

mixing device]. This agreement extends as close to the inlet as could be measured where the field was not yet uniform on a coarse scale (radial average values had to be made in this region).

Additional direct measurements of X and d or I_s by McKelvey, *et al.* (1974) add further confirmation to Eqs. (26) through (29). Thus, it appears that we have working equations which, for very rapid reactions, allow prediction of the mean conversion, given the mixing decay function d. (It might be noted that this gives a relatively simple way to measure mixing, d, since mean quantities rather than r.m.s. quantities are measured.)

Although the method has necessarily been roundabout, what we have effectively done is solved the design equation for a one-dimensional reactor

$$dX/dZ = -N_D[X(X + \beta - 1)/\beta - \psi^2] \tag{32}$$

which is the one-dimensional form of Eqs. (6) and (3) with axial dispersion neglected, and the condition (to be discussed later),

$$\overline{ab}_0 = -\overline{A}_0\overline{B}_0 \tag{32a}$$

(The solution does include the effect of axial dispersion, but it does not appear to be important.) In Eq. (32), the characteristic mixing length, L_m, is used. This is the distance required for d to fall to some given value, say 1/2.

It is seen from Eq. (32) that the very rapid reaction (the diffusion controlled limit) occurs when N_D, the first Damköhler Number is large and this limit depends upon β and ψ. At the limit

$$X(X + \beta - 1)/\beta = \psi^2 \tag{33}$$

and for $\beta = 1$

$$X = \psi \tag{34}$$

and our solution, Eqs. (26) and (28), from this viewpoint can be considered to give ψ as a function of the mixing parameter d. This is implicit in Eq. (26) and explicit in eq. (29) from which we obtain

$$\psi = d \tag{35}$$

for stoichiometric mixtures.*

Thus, for slower speed reactions (N_D not large) the problem is to evaluate ψ. From Eqs. (4) and (5)

$$\psi = \psi(Z, N_D, \beta, N_{Re}, w_A/w_B, N_{Sc}) \tag{36}$$

We have left out the diffusivity ratio which is considered unimportant. As before, we cannot attack this problem directly but turn to the other limit, corresponding to the very slow reaction considered by Toor (1969).

---

*A subsidiary result for the stoichiometric reaction is (see appendix):

$$(n^2 a'^2 + b'^2)/(n^2 a_0'^2 + b_0'^2) = d^2 = I_s \tag{A-4}$$

O'Brien (1971) has shown that if species A and B are identically distributed initially then

$$\overline{A^2} = \overline{B^2} = (1/2)\overline{J^2}$$

V. VERY SLOW REACTION

The very slow reaction is defined as one where

$$\Psi^2 << X(X + \beta - 1)/\beta \qquad (37)$$

so the reaction is effectively homogeneous and the limit is the case in which species A and B do not react at all. Thus if, as before, species A enters through the 1 jets and species B enters through the 2 jets, we write

$$(A - 0)/(A_1 - 0) = Y_A \qquad (38)$$

$$(B - B_2)/(0 - B_2) = Y_B \qquad (39)$$

Then Y_A and Y_B solve Eq. (12) (diffusivities of A and B are assumed equal as before) with the boundary conditions

$$\begin{aligned} &\text{at the 1 inlet} \qquad Y_A = Y_B = 1 \\ &\text{at the 2 inlet} \qquad Y_A = Y_B = 0 \end{aligned} \qquad (40)$$

and on all other boundaries

$$\partial Y_A/\partial y = \partial Y_B/\partial y = 0 \qquad (41)$$

here y is the distance normal to the wall. Hence

$$Y_A = Y_B \qquad (42)$$

and from Eqs. (37) and (38)

$$A = -(A_1/B_2)(B - B_2) \qquad (43)$$

Then

$$a = -(A_1/B_2)b \qquad (44)$$

so

$$a' = (A_1/B_2)b' \qquad (45)$$

Furthermore, multiplying Eq. (44) by b, averaging and using Eq. (45),

$$\overline{ab} = -(A_1/B_2)b'^2 = -(B_2/A_1)a'^2 \tag{46}$$

Hence, for the one-dimensional reactor, dividing Eq. (46) by its value at the inlet,

$$\psi = d_A = d_B = d \tag{47}$$

which is identical to the result obtained for very rapid reactions of stoichiometric mixtures (but not the same for non-stoichiometric mixtures).

VI. THE INLET CONDITION

Equation (32a) is clearly valid for an infinitely rapid reaction. It also follows for a very slow reaction from Eq. (46) at Z = 0 and Keeler, *et al*'s(1965) calculation of a'_0 for an ideal inlet,

$$a'_0 = \overline{A}_0(w_B/w_A) \tag{48}$$

and the material balances

$$A_1 = (w/w_A)\overline{A}_0 \tag{49}$$

$$B_2 = (w/w_B)\overline{B}_0 \tag{50}$$

In fact, Eq. (32a) must be true for all reaction speeds since the concentration fields at an ideal inlet (no back mixing) cannot be effected by the reaction, so the equation must be true for all values of N_D.

VII. THE INVARIANCE HYPOTHESIS

Equations (35) and (47) can be summarized as follows:

$$\psi(Z,\infty,1,N_{Re},w_A/w_B,N_{Sc}) = \psi(Z,0,\beta,N_{Re},w_A/w_B,N_{Sc}) \quad (51)$$
$$= d(Z,N_{Re},w_A/w_B,N_{Sc})$$

Thus, for the stoichiometric reaction, the two limits of Damköhler Number give precisely the same result, which led Toor (1969) to postulate that this invariance held over all the intermediate values of N_D,

$$\psi(Z,N_D,1,N_{Re},N_{Sc}) = d(Z,N_{Re},N_{Sc}) \quad (52)$$

If Eq. (52) is valid, then Eq. (32) becomes a useful design equation (for stoichiometric mixtures).* Toor (1969) attempted to test this with the earlier data of Vassilatos and Toor (1965), but the results were inconclusive.

Mao and Toor (1971) then set out to test this hypothesis using the improved mixing device of Shuck (1969). This device used 188 close packed tubes and experiments of McKelvey *et al.* (1974) showed that the gross velocity field in this device is more uniform than in the device of Vassilatos.

---

*A similar comparison of the fluctuations of the *individual* reactants is possible with Eq. (A-4). Since for the very slow reaction

$$a'/a_0' = b'/b_0' = d$$

identical behavior at the two limits is not obtained unless $na' = b'$. One case where this would be true would be that in which $n = 1$, flow rates of reactants are equal and the mixing device is symmetric (as in the multiple jet devices). Because $\beta = 1$, these strict conditions by symmetry require $a' = b'$ and hence Eq. (A-4) for a very rapid reaction reduces to the above equation for a very slow reaction.

Mao and Toor (1971) used the reaction systems described in Table 2. The three most rapid reactions with

$$1.4 \times 10^{7} \frac{\ell}{\text{mole sec}} \leq k \leq 1.4 \times 10^{11} \frac{\ell}{\text{mole sec}}$$

TABLE 2 *Reaction Systems Used by Mao and Toor*

System	Rate constant k, ℓ/mole sec	Ionic Strength
NCl-NaOH	1.4×10^{11}	0
Maleic Acid $(II)^{a}$-OH^{-}	3.0×10^{8}	0.1 (KNO_3)
Nitrilotriacetic Acid $(III)^{b}$-OH^{-}	1.4×10^{7}	0.1 (KNO_3)
CO_2-2NaOH	8.32×10^{3}	0

[a] The second stage acid group of maleic acid.

[b] The third stage acid group of nitrilotriacetic acid.

all gave identical results further confirming the concept of a diffusion controlled limit and the insensitivity of this limit to both molecular diffusivity and diffusivity ratio. They used two Reynolds Numbers and a typical run is shown in Fig. 6. Eq. (29) was then used to determine d and the results are shown in Fig. 7 and 8 for two different values of N_{Re}.

The data for the reaction of CO_2 with NaOH are shown in Fig. 9 where the two limits, homogeneous and diffusion controlled, are shown. Data corrected for the presence of HCO_3^- as well as uncorrected data are given.

Since stoichiometric mixtures were used, Eq. (32) becomes

$$dX/dZ = -N_D[X^2 - \psi^2] \tag{53}$$

and ψ^2 was determined from the measurements by means of this equation.

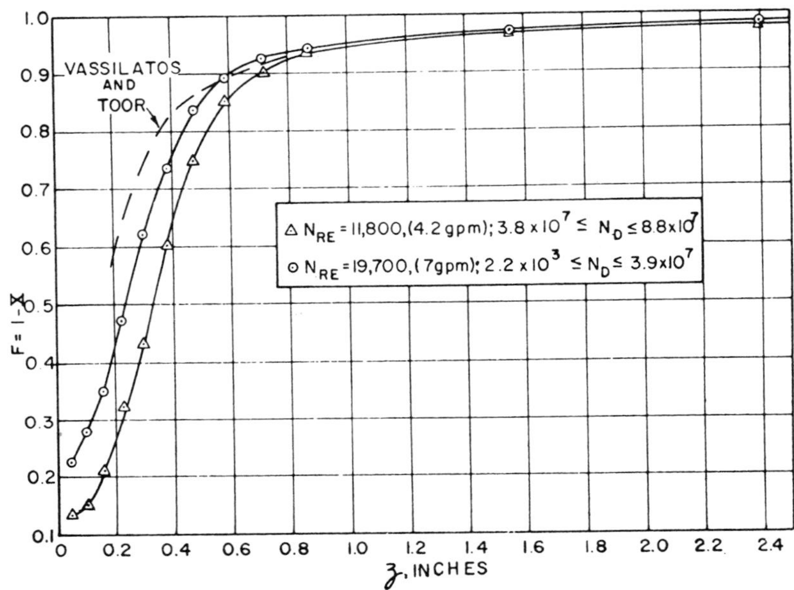

Fig. 6. Conversion vs. distance, very rapid reactions. From Mao and Toor (1971).

In Figs. 7 and 8, d^2 and ψ^2 are compared at the two Reynolds Numbers used. ψ^2 is consistently greater than d^2 but the differences are small enough to be an artifact. The assumption that $\psi = d$ was used with Eq. (53) to compute the curve shown in Fig. 9. There is some error at the greater lengths, but it is not large. Thus, the hypothesis embodied in Eq. (52) appears to be at least a useful approximation; a theoretical calculation of ψ remains as an interesting challenge to the statistical theory of turbulence.

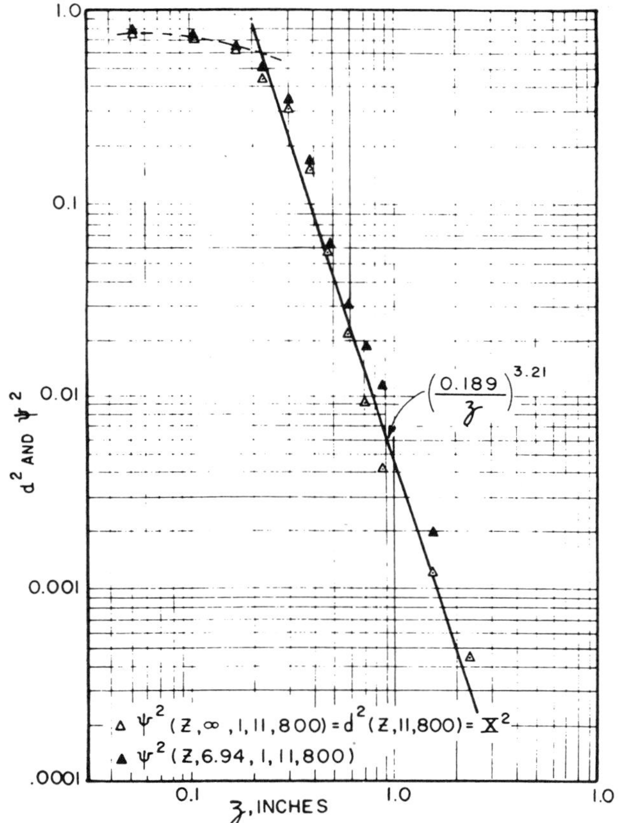

Fig. 7. I_s or d^2 and ψ^2 (same as Ψ) at 4.2 gpm. From Mao and Toor (1971).

VIII. NON-STOICHIOMETRIC REACTIONS

Since the statistical model of very rapid reactions has been confirmed for values of β up to 10, the invariance hypothesis can be shown to be incorrect for non-stoichiometric reactions since it diverges increasingly from the statistical model as β increases. However, Mao and Toor (1971) showed that, for values of β of 1.5 and 3.0, the hypothesis was

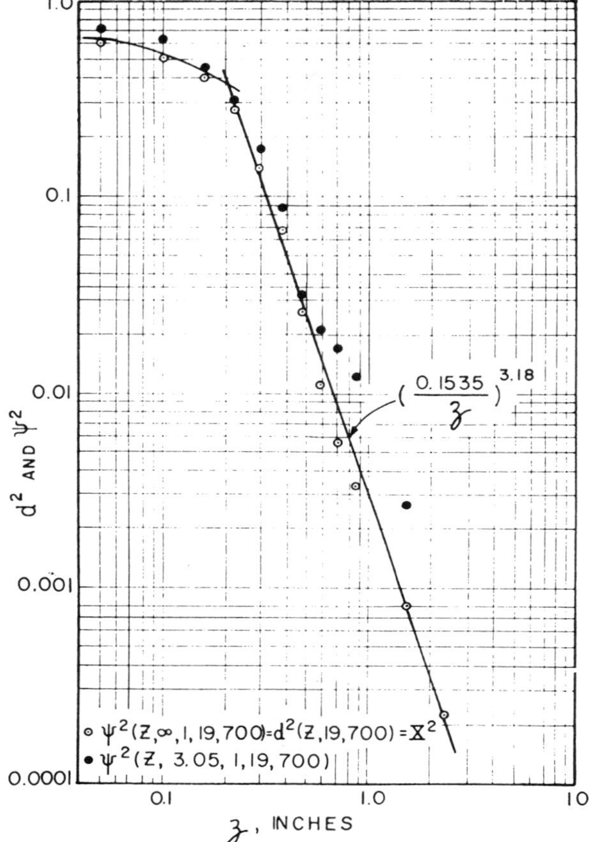

Fig. 8. I_s or d^2 and ψ^2 (same as Ψ) at 7.0 gpm. From Mao and Toor (1971).

reasonably good [replace ψ^2 by d^2 in Eq. (32)] as shown in Fig. 10. It would be expected that this approach would fail for higher values of β and here there is need for a more complete theory.

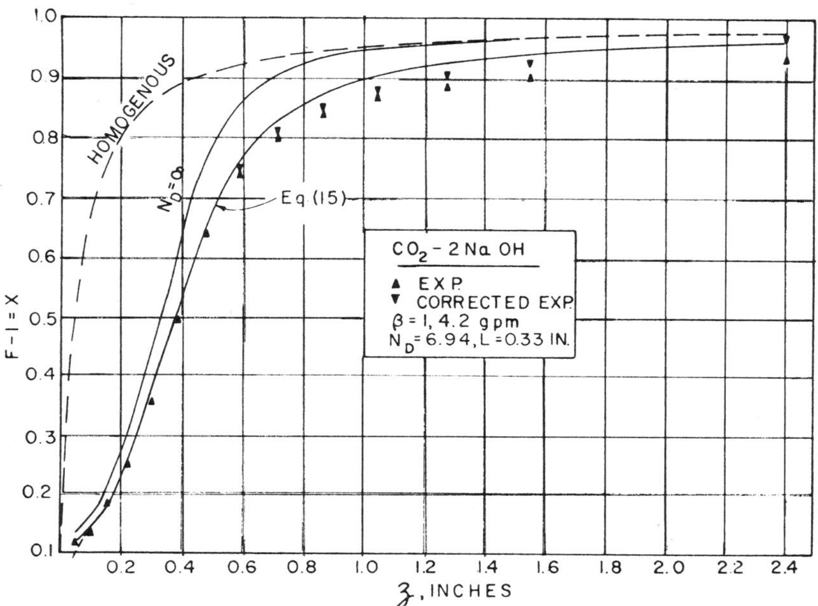

Fig. 9. Rapid stoichiometric reaction at 4.2 gpm. Equation (15) is Eq. (53) with $\psi = d$. From Mao and Toor (1971).

IX. INLET REGION

Mao and Toor (1971) were able to measure conversion to within 0.05 in. of the inlet and, even at this point, significant conversion has taken place (see Figs. 6, 9 and 10). In this region, the system is not truly one-dimensional for the jets have not lost their identity and the measured conversion should be considered to be some radial average value. Vassilatos and Toor (1965) showed, however, that for very rapid reactions, it was satisfactory to treat the system as if it were one-dimensional right from the inlet.

In order to handle this region with intermediate speed reactions, Mao and Toor (1971) used a stirred tank

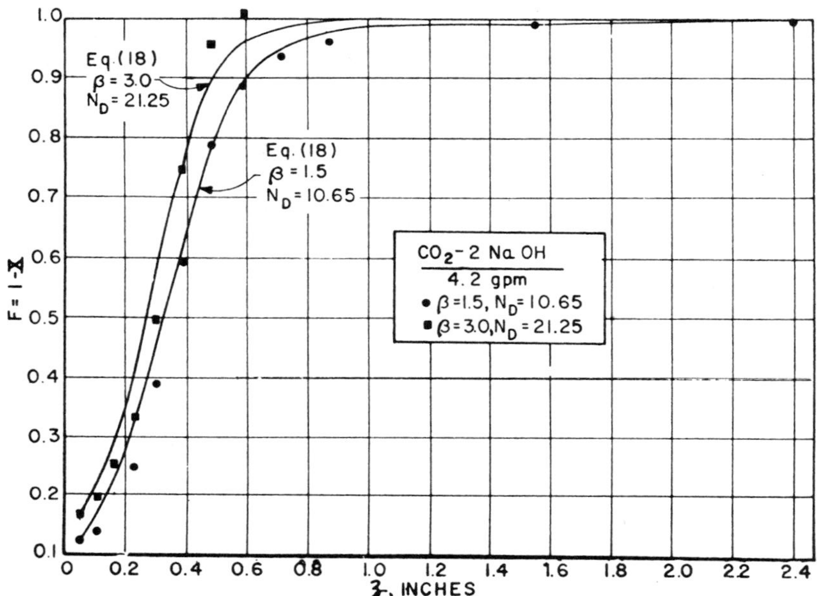

Fig. 10. Nonstoichiometric rapid reactions at 4.2 gpm. Equation (18) is Eq. (32) with $\psi = d$. From Mao and Toor (1971).

model, but McKelvey, *et al.*(1974) show that one can dispense with this model and integrate Eq. (53) from the inlet (Z = 0).

X. GAS PHASE REACTIONS

All the above studies were carried out in liquid (aqueous) systems and although they showed no effect of diffusivity (N_{Sc}) or diffusivity ratio, the range of diffusivities possible is too narrow to allow extrapolation to gases.

In order to investigate gases, Ajmera (1969) attempted to find a suitable gaseous reaction which would be diffusion controlled. The fastest suitable reaction he could find was the reaction

$$NO + O_3 \rightarrow NO_2 + O_2$$

Because of the small quantities of O_3 available, he found it necessary to go to small reactors and the two of interest here were

Reactor B: 14 close packed tubes, 0.028" i.d. in a 0.125" i.d. tube. Reactants in alternate tubes.

Reactor C: A 0.095" tube located concentrically in a 0.125" i.d. tube.

He carried out the study much as before, diluting the reactants with N_2 and Freon 13, and following the temperature rise along the reactor. Since the rise was as much as $40^{o}C$, he had to account for the temperature dependence of k (but fluctuations of k were not important).

He found reactor B to be reaction controlled - in fact, his measurements extended the known temperature range of the reaction velocity constant.

Reactor C was neither reaction nor diffusion controlled, so he used Eq. (53) to extract ψ^2 which is shown in Figs. 11,12 and 13.

Since this reactor differs from those used for liquids, Singh (1973) subsequently ran the very rapid aqueous phase, HCl-NaOH stoichiometric reaction in the same reactor and his data are compared to Ajmera's in Figs. 11, 12 and 13. The results are the same within experimental error and we conclude that:

(1) d is essentially the same for gases and liquids at the same Reynolds Number. The effect of Schmidt number is negligible.

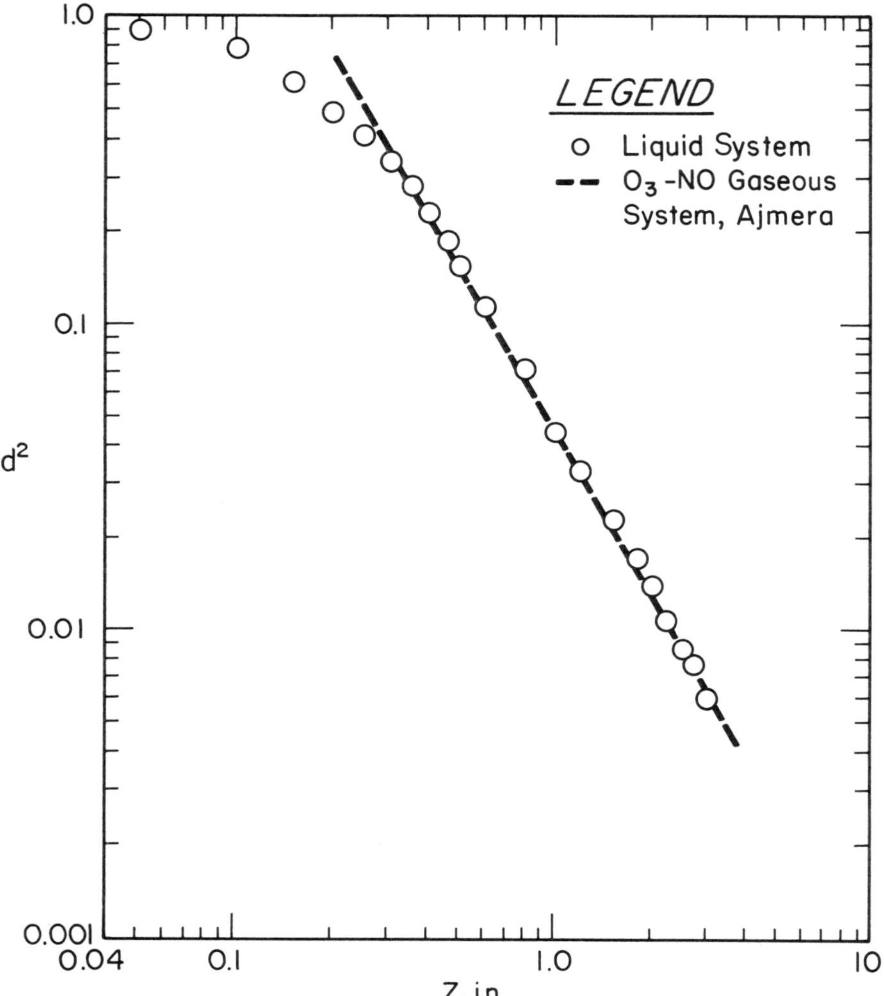

Fig. 11. Comparison of decay laws in liquid and gaseous systems, reactor C, N_{Re} = 3,500. From Singh (1973).

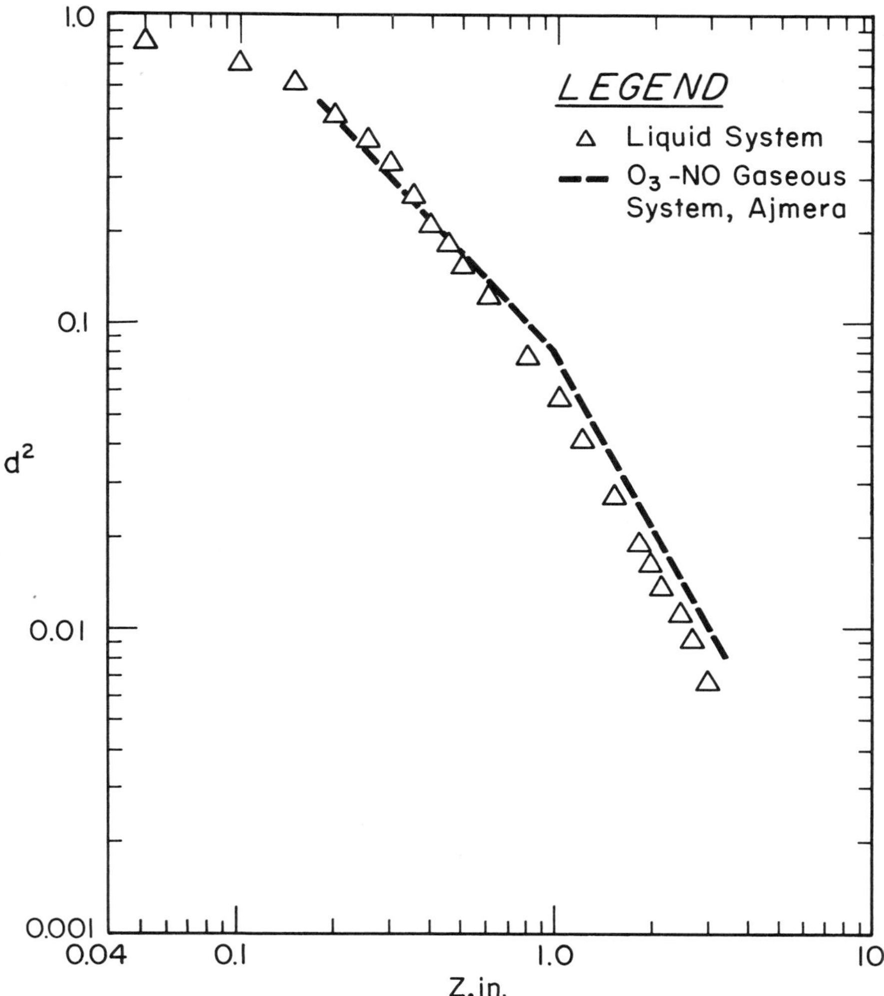

Fig. 12. Comparison of decay laws in liquid and gaseous systems, reactor C, $N_{Re} = 6{,}700$. *From Singh (1973).*

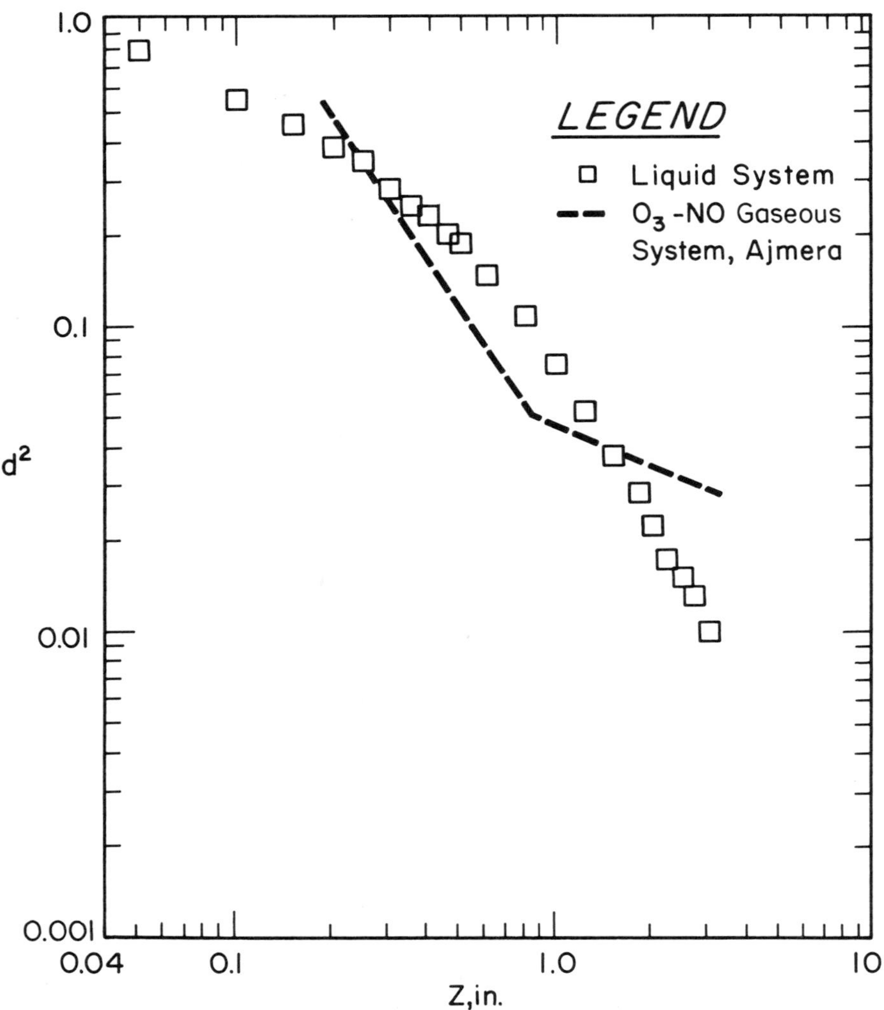

Fig. 13. Comparison of decay laws in liquid and gaseous systems, reactor C, N_{Re} = 12,000. From Singh (1973).

(2) The invariance hypothesis applies to gases as well as liquids, at least to within the accuracy of these experiments.

This equality between gases and liquids is most convenient since it allows the use of aqueous acid-base reactions for the determination of mixing (d) for gaseous systems. In effect, gas mixing can be predicted as pointed out in Chapt. II.

The insensitivity of ψ (or d) to molecular diffusivity seems paradoxical, particularly since we talk about a diffusion controlled limit and reaction surfaces and we are confident that in all cases the final step in the mixing process must be a molecular diffusion process. This paradox becomes less disturbing if one recognizes that on the fine scale molecular diffusion is very rapid; the diffusion relaxation time θ_d is quadratic with scale,

$$\theta_d = K\ell^2/D$$

where ℓ is a suitable scale such as the size of the small eddies and K is order one. Hence, the qualitative picture must be one in which the molecular diffusion plays no significant role until the turbulent mixing reduces the initial large scale of reactants A and B to one fine enough for the diffusion to proceed very rapidly. Presumably, the time (or distance along the reactor) for the mixing *step* is large compared to the time (or distance) for the molecular diffusion *step*.

XI. DESIGN

If the invariance hypothesis is accepted as sufficiently accurate for practical design, then at least for $\beta < 3.0$, Eq. (31) can be written as

$$dX/dZ = -N_D[X(X + \beta - 1)/\beta - d^2(Z, N_{Re}, w_A/w_B)] \quad (54)$$

If d is known, the characteristic mixing distance L_m which appears in Z and N_D is also known and Eq. (54) can be integrated over the full range of Z for all values of N_D. A mapping of this sort from Mao and Toor (1971) is shown in Fig. 14 in terms of fractional conversion. This is for fixed

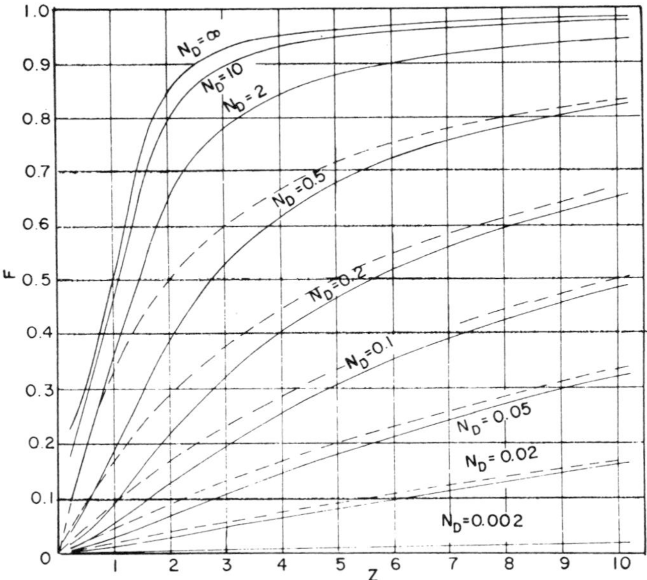

Fig. 14. Computed values of $F(Z, N_D)$ at 7 gpm. Dotted lines are homogeneous conversion. From Mao and Toor (1971).

values of β, N_{Re} and w_A/w_B and for a given mixing device.

The decay law d may be obtained either by direct measurement of concentration fluctuations, using a tracer, or by measuring X with a very rapid reaction (or less desirably, an intermediate speed reaction) or from theory. The simplest experimental method is to use a very rapid, stoichiometric, acid-base reaction in water where X = d.

In practical situations, of course, one would like to determine d in a pilot reactor. In principle, this can be done by measuring d in a pilot reactor which is geometrically similar to the full-scale reactor, but there may be technical difficulties in covering the Reynolds Number range desired.

If one could operate a pilot reactor at the same Reynolds Number as the full-scale reactor (Toor and Singh, 1973), the Damköhler Number in the two cases are related by

$$N_{D_{II}} = N_{D_I} (L_{II}/L_I)^2 \tag{55}$$

so that mixing effects are likely to be more important in the larger reactor.

This is seen by examination of Fig. 14. Below a lower critical value of N_D (the value depending upon how much error is tolerable), the system is reaction controlled, mixing effects are negligible and the homogeneous rate law applies, while above some upper critical value, the system is diffusion controlled.

It is convenient to look at this in terms of reaction and mixing times,

$$N_D = \theta_m/\theta_r \tag{56}$$

The reaction half-life is independent of scale but the mixing half-life at a *constant Reynolds Number* varies as

$$\theta_{m_{II}} = \theta_{m_I} (L_{II}/L_I)^2 \tag{57}$$

In a small multiple jet reactor (Ajmera's Reactor B scaled by a factor of 2),

$$\theta_m = 0.01 \text{ sec.}$$

so, in this reactor, mixing effects are very small for reaction half-lives less than 0.5 sec. and not very large even at 0.05 sec. This suggests that mixing is rapid enough compared to the speed of most industrial reactions so that mixing effects can be mostly ignored, but both the effects of scale and of mixing device design need to be considered.

For example, if the device considered is scaled by a factor of 30 to give an 8-inch diameter reactor and the Reynolds Number is held constant, then from Eq. (57) θ_m rises to 9 seconds and the above reaction half-lives rise to 450 and 45 seconds, respectively. But this is still inconclusive, for the relatively low N_{Re} used in the small reactor is not likely to be used in the large reactor - data for d as a function of N_{Re} is needed and only limited data are available. Figures 15 and 16 give a few examples. Note that jet Reynolds Number is the appropriate Reynolds Number in the jet mixers (see below). They do suggest that if the velocity is held constant or increased upon scaling up, then the effect of scale upon N_D can be decreased or eliminated and hence, most industrial reactors with well-designed mixing heads probably can be operated in the reaction controlled region. More data on the behavior of d is needed, however, before a definitive answer is possible.

XII. OTHER MODELS

Other approaches to predicting intermediate speed reactions exist. For example, Mao and Toor (1970) used a diffusion model with alternate slabs of reactants which were allowed to interdiffuse and react. The geometry of the slabs was chosen to fit the data for very rapid stoichiometric reactions and then this geometry was used to predict conversions for slower speed reactions and different stoichiometric

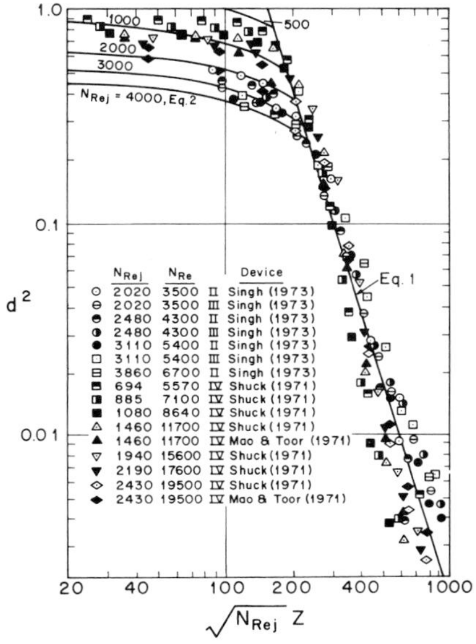

Fig. 15. I_s or d^2 vs. $\sqrt{N_{Rej}}\,Z$, multiple jet devices. For equations see references. From Singh and Toor (1973).

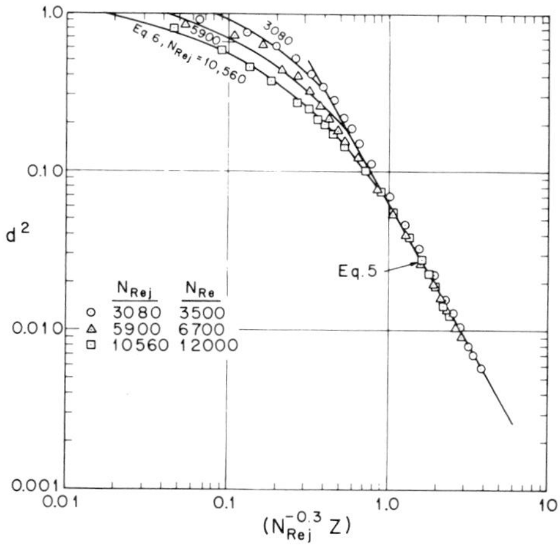

Fig. 16. I_s or d^2 vs. $N_{Rej}^{-0.3}Z$, single jet device. For equations see references. From Singh and Toor (1973).

ratios. A typical result is shown in Fig. 17.

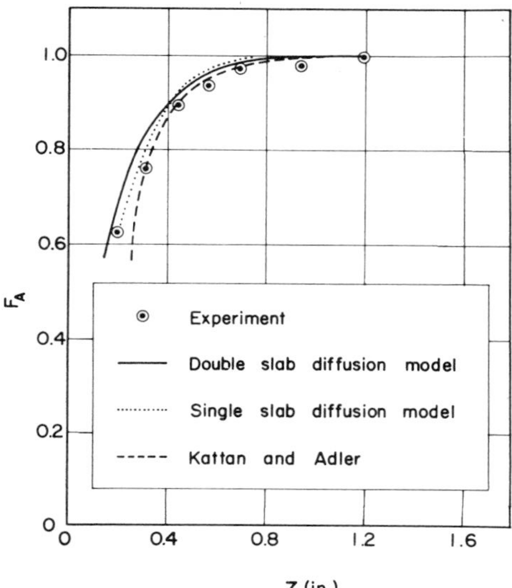

Fig. 17. Fractional conversion ($F_A = F$) vs. distance, k = 12,400 liter (mole) (sec.), β = 2.52. From Mao and Toor (1970).

Kattan and Adler (1967) and others (discussed in more detail in Chapt. V) have developed random coalescence-dispersion models which give similar results. All these models more or less fit the Vassilatos and Toor (1965) data (see Fig. 17).

None of these models have been compared to the intermediate speed data of Mao and Toor (1971) which is believed to be more reliable than the data of Vassilatos and Toor (1965), both because of the improved mixing device and improved techniques used by Mao and Toor. However, the existing work suggests that most, if not all, of the models might be used to predict the intermediate speed reactions.

The model of Mao and Toor based on the invariance hypothesis is the simplest to use and has the most physical content, but it is limited to reactions not too far from stoichiometric and to a specific reaction scheme while the other models can, in principle, handle all ranges of stoichiometry and more complex reaction schemes (although this has not been confirmed experimentally). For more complex systems, detailed modeling would appear to be necessary; thus, attention is devoted to this in Chapts. IV and V.

XIII. TWO-DIMENSIONAL REACTOR

Singh (1973) has studied the two-dimensional reaction-mixing problem by using a 3/8 inch jet in a 3 inch i. d. pipe. One reactant is in the jet and one in the slow moving outer stream. This is close to an unconfined jet since measurements were made before the expanded jet reached the wall. Singh studied very rapid acid-base reactions as well as the CO_2-2NaO reaction used to give intermediate speed data in the one-dimensional experiments. He showed that by use of two independent experiments with suitable inlet temperature differences, one could determine time average concentrations of reactants from measurements of the temperature fields.

Equation (19) predicts the equivalence of $\overline{Y}$ and $\overline{Y}_m$ and, if molecular diffusivity plays no role, the equivalence of $\overline{Y}$ and the reduced temperature profile $(\overline{T} - T_2)/(T_1 - T_2)$. Figure 18 shows that this is indeed the case.

The statistical model which has been confirmed in one-dimensional systems is tested in Fig. 19 which compares experimental data for diffusion controlled acid-base systems ten diameters downstream with Eq. (24), etc. The measured values of γ_m obtained by tracer experiments in an unconfined jet (Lawrence, 1965), and the measured values of $\overline{Y}_J$ were used here. Since the effect of the reaction is very large, this

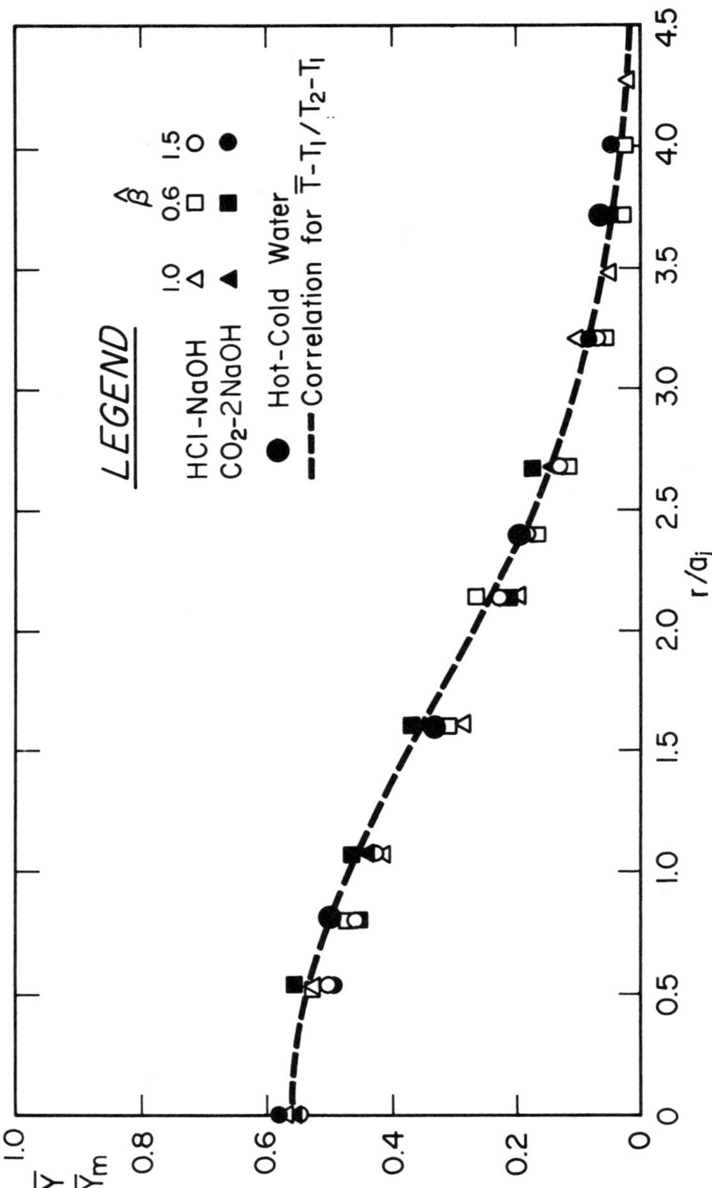

Fig. 18. $\bar{Y}$, $\bar{Y}_m$ *and* $(\bar{T} - T_1)/(T_2 - T_1)$ *vs.* z/a_j. *From Singh (1973).*

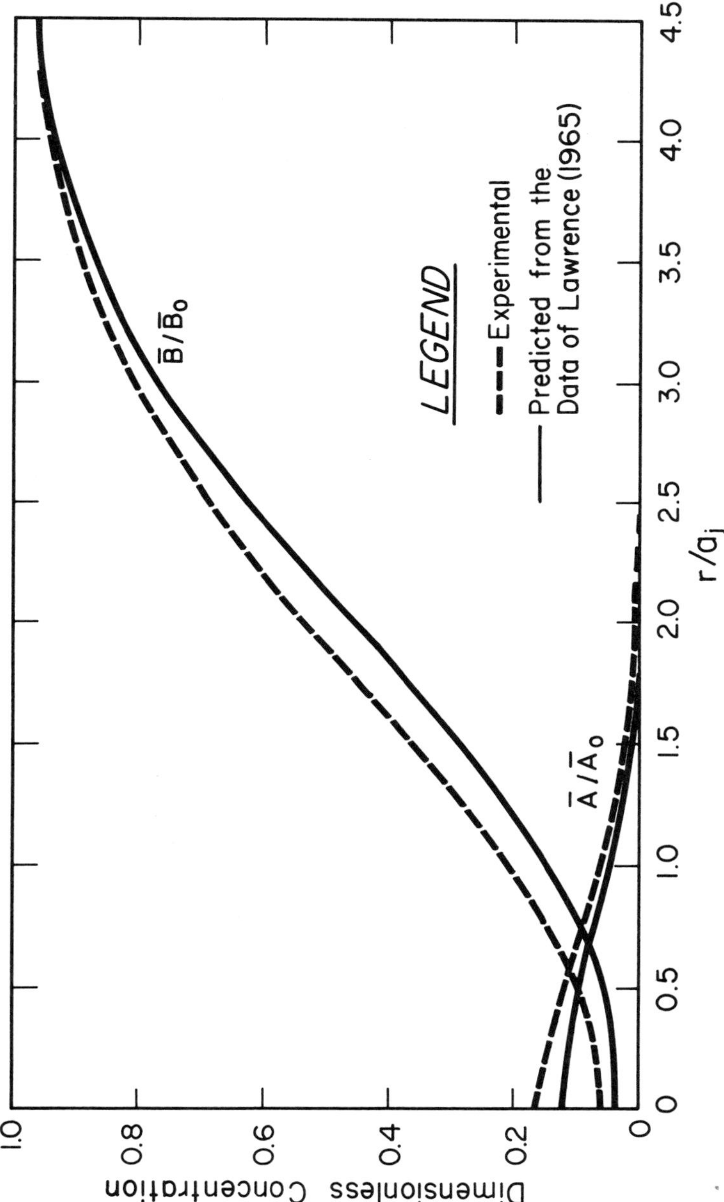

Fig. 19. $\bar{A}/\bar{A}_0$ *and* $\bar{B}/\bar{B}_0$ *at* z/a_j *using* γ_m *data of Lawrence. From Singh (1973).*

is a significant test of the model and the results are considered to be in reasonably good agreement with the model.

There is no result analogous to Eq. (51) in multidimensional systems - except along a surface at which $\overline{J}$ is constant. Singh (1973) assumed, but with less confidence than in the one-dimensional case,

$$(\overline{ab}/A_0B_0)_k = (\overline{ab}/A_0B_0)_{k = \infty}$$

That is, that the mean of the reactant product fluctuations is independent of the reaction velocity constant (or appropriate Damköhler Number).

Thus he used the very rapid reaction data to predict $\overline{ab}$ for the slower speed CO_2-2NaOH reaction. Then, numerically integrating Eqs. (6) and (7), he obtained the results shown in Fig. 20 which also contains experimental data for this reaction.

These results are really inconclusive even though the predictions are not bad, for he found that in his experiments, this reaction was close to homogeneous and hence conversion is quite insensitive to $\overline{ab}$.

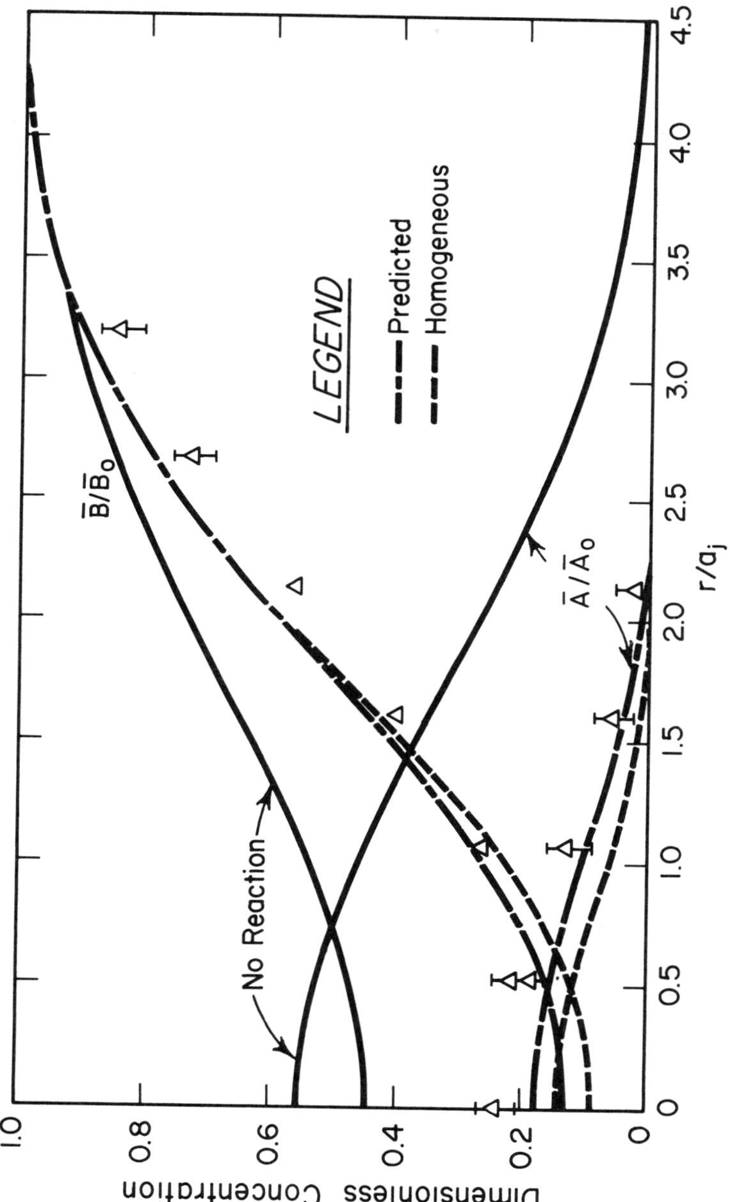

Fig. 20. Prediction of Reaction-mixing model, $z/a_j = 9.82$. From Singh (1973).

XIV. SUMMARY

In summary, in the absence of a tractable statistical theory of turbulence, ad-hoc methods and experiment have allowed the development of enough understanding of the process of turbulent mixing with the reaction

$$A + B \rightarrow \text{Products}$$

to permit rational design in non-premixed tubular reactors and some progress has been made in understanding multidimensional systems.

More complex reaction systems and high non-isothermal systems remain as challenging problems.

XV. APPENDIX

DERIVATION OF EQ. (A-4)

From Eq. (8a)

$$j = na - b \tag{A-1}$$

so

$$\overline{j'^2} = n^2\overline{a'^2} - 2n\overline{ab} + \overline{b'^2} \tag{A-2}$$

Substituting from Eqs. (18), (20) and (29) and using the stoichiometric condition $\overline{A} = \overline{B}$

$$d^2 = (n^2\overline{a'^2} + \overline{b'^2})/(\overline{j_0'^2} + 2n\overline{a_0'^2}) \tag{A-3}$$

From Eqs. (A-2) and (20) and the stoichiometric condition

$$\overline{j_0'^2} = n^2\overline{a_0'^2} - 2n\overline{A}_0^2 + \overline{b_0'^2}$$

so Eq. (A-3) becomes

$$d^2 = (n^2\overline{a'^2} + \overline{b'^2})/(n^2\overline{a_0'^2} + \overline{b_0'^2}) \tag{A-4}$$

XVI. REFERENCES

Ajmera, P.V. (1969) Ph.D. Thesis, Carnegie-Mellon University.

Burke, S.P. and Schumann, T.E.W. (1928) *Ind. Eng. Chem. 20,* 998.

Hawthorne, W.R., Weddell, D.S. and Hottel, H.C. (1949) THIRD SYMPOSIUM ON COMBUSTION FLAME AND EXPLOSION, p. 266, Williams and Wilkins Co., Baltimore, Md.

Kattan, A. and Adler, R.J. (1967) *A.I.Ch.E.J. 13,* 580.

Keeler, R.N., Petersen, E.E. and Prausnitz, J.M. (1965) *A.I. Ch.E.J. 11,* 221.

Lawrence, W.J. (1965) Ph.D. Thesis, University of California at Berkeley.

Mao, K.W. and Toor, H.L. (1970) *A.I.Ch.E.J. 16,* 49.

Mao, K.W. and Toor, H.L. (1971) *I & EC Fund. 10,* 192.

McKelvey, K.N., Yieh, H.-N., Zakanycz, S. and Brodkey, R.S. (1974) to be published.

O'Brien, E.E. (1971) *Phys. of Fluids 14,* 1326.

Shuck, D.L. (1969) Carnegie-Mellon University, private communication.

Singh, M. (1973) Ph.D. Thesis, Carnegie-Mellon University.

Toor, H.L. (1962) *A.I.Ch.E.J. 8,* 70.

Toor, H.L. (1969) *I & EC Fund. 8,* 655.

Toor, H.L. and Singh, M. (1974) *A.I.Ch.E.J.,* to be published.

Vassilatos, G. and Toor, H.L. (1965) *A.I.Ch.E.J. 11,* 666.

Turbulent Mixing in Chemically Reactive Flows

S. N. B. Murthy

Chapter IV

Turbulent Mixing in Chemically Reactive Flows

S. N. B. MURTHY

School of Mechanical Engineering
Purdue University
West Lafayette, Indiana 47907

I. ABSTRACT

The principal feature of chemical flows is the presence of scalar quantities. While the understanding of the structure of the mixing process (especially the relation between the strong, coherent, large scale structure and the ultimate mixing at the molecular level) is largely ambiguous and controversial, considerable advances have been made in the modeling of mixing layers. Both deterministic and statistical continuum theories and models have been developed. However, it appears that correlations and spectral analysis are inadequate for a detailed understanding of mixing processes involving scalar quantities.

II. INTRODUCTION

The problem of reactive mixing of shear flows is of considerable technological importance in combustion, jet propulsion, chemical lasers and environmental studies. While turbulent mixing continues to be largely an intractable subject, there exists a considerable need for the development of at least *ad hoc* methods for the calculation of mean flow properties. A number of predictive methods have therefore been developed, based largely on heuristic reasoning. Improvements in such methods have arisen principally through the use of large scale computational facilities (Launder and

and Spalding, 1972).

Any basic developments in turbulent mixing will arise, of course, only through experiments. One wishes that the importance of the subject would lead to extensive measurements in the various flow fields. Progress in experiments requires some agreement on the following points: 1) what parameters are of significance in mixing, 2) what quantities should be measured, 3) what is the accuracy of the possible measurement and sampling techniques and 4) what mixing configurations best permit a unified picture of turbulent mixing.

The ideal approach would be for two or three definitive flow configurations to be selected for intensive study from several points of view and using different measurement techniques. For example, the study of a chemically reactive flow requires the development of a flow configuration with definable nonequilibrium characteristics and with controllable concentration and reaction gradients in addition to properly specified initial conditions. The measurement techniques as well as the analogue and digital processing of data have improved steadily over the past decade. However, there are limitations that must be considered in the methods currently being used, namely, hot-wire anemometry, laser-Doppler velocimetry, holography and Raman and Rayleigh scattering techniques. The analysis of data through conditional sampling and image processing is a subject of current research. Finally, physical experiments should be supplemented by computer experiments.

The principal objective here is to discuss some of the recent approaches to the problem of nonhomogeneous mixing involving multiple reactive species and heat. It is hoped that this will provide a background to the further development of some of the more fruitful approaches to the problem of turbulent mixing. We restrict attention throughout to high Reynolds number flows away from physical boundaries.

After recalling briefly some of the experimental results on turbulent mixing of shear flows, we shall discuss modeling of turbulent flows which include transport and chemical reaction. This will be followed by a summary of statistical continuum theories for reactive flows. In both cases, the general approach is to examine how chemical reaction may be superimposed on turbulent mixing.

Finally, in view of the great interest in turbulent flames, some recent advances in turbulent diffusion flame theory are presented in some detail.

Details regarding many aspects of the problems discussed here may be found in the Proceedings of the SQUID Workshop on Turbulent Mixing of Non-Reactive and Reactive Systems (Murthy, 1974).

III. EXPERIMENTS ON SHEAR LAYER MIXING

A simple flow configuration which illustrates the problem of the mixing of shear layers consists of two flows (built up on either side of a flat plate of negligible thickness) which come into contact downstream of the trailing edge of the plate. In general, the problem is to describe the mixing taking place between the two streams under various conditions when turbulence exists. In chemically reactive flows, the problem of interest is when each stream consists of multi-component fluids so that the mixing between molecular species can be examined. The variables in such a configuration are 1) the number of species including active and inert constituents, 2) the rate of reaction 3) the order of the reaction, 4) the reversibility of the reaction, 5) the proportion of the species in relation to stoichiometry, 6) the thermic nature of the reaction and 7) the molecular diffusion characteristics of the different species. Such flows are nonlinear in nature not only because of the character of the equations governing the stochastic variables but also because of the inevitable coupling between the velocity and the scalar fields.

In experimental studies on such flows, it is common practice to measure the following: 1) the spreading rate of the mixing layer or the entrainment rate, 2) the mean velocity and pressure profiles, 3) the structure of the mixing layer turbulence, 4) the turbulent shear stress distribution and 5) the intermittency profiles (Abramovich, 1963; Bradshaw, 1971). When chemical reaction is involved, it is usual to consider the following additional parameters: 1) reaction zone thickness and spread, 2) volumetric heat release and 3) product distribution of the mixture. Thus the mixing process in reactive flows can be considered as a two-stage process, entrainment which may be looked upon as *engulfment* (Brown and Roshko, 1974) and molecular mixing which will ultimately lead to chemical action. Thus, in the splitter plate experiments of Brown and Roshko (1971) using helium and nitrogen, the density data obtained from the aspirating probe (Brown and Rebello, 1972) appear to show that while each gas penetrates into the region of the other gas, there is little molecular mixing and an interface separates the two gases. It must not, of course, be forgotten that the interface is highly convoluted and viscosity and diffusivity will tend to smear the region. This observation is of fundamental implication for the entrainment process, especially in reactive systems. While the convoluted interface clearly points to the existence and importance of the large coherent structure in turbulent mixing, it appears difficult to show a direct connection between such coherent structure and the entrainment process as opposed to the influence of the behavior of the turbulent-non-turbulent interface on entrainment.

The influence of Reynolds number, Mach number, velocity ratio and density ratio on the spreading parameter of a mixing layer has been discussed in detail by Birch and Eggers (1972). In regard to compressibility effects, it is not yet possible to distinquish between the effect of density differences arising on account of concentration, velocity or temperature differences between the two streams (Libby, 1973). We shall return to this subject in Section IV.

III.A. *SOME EXPERIMENTAL RESULTS*

We shall summarize here a number of experimental findings on the turbulent mixing of shear flows.

The first detailed measurements were made (Liepmann and Laufer, 1949) in a mixing layer developed between a two-dimensional jet of air and the still air adjoining it on one of its boundaries. A class of experiments has come into existence based on modifications to the original flow configuration. The note-worthy among them, which deal with momentum mixing in the absence of scalar quantities, are due to Wygnanski and Fiedler (1970), Spencer and Jones (1971), Patel (1973), Brown and Roshko (1971) and Winant and Browand (1974). It may be pointed out that Brown and Roshko did examine the mixing of helium and nitrogen streams but since the concentration of the two individual species were not determined in the mixing layer, the experiments should be considered as dealing with the influence of combinations of density and velocity differences between the mixing streams.

Experiments which involve scalar quantities in the mixing layer are due to Shackleford *et al.* (1973), Grant *et al.* (1973), Graham *et al.* (1975), Stanford and Libby (1974) and La Rue and Libby (1974). They are of fundamental importance in chemically reactive flows.

Shackleford *et al.* have reported on the mixing zone between parallel, supersonic (Mach number ≃ 4) streams of hydrogen and florine. The temperatures of the two streams were controlled independently by electric arc heating. An I. R. scanner was used to obtain the infrared emission from the reaction zone and, in combination with a source capable of emitting lines with known spectral contours, both the concentration of HF in the vibrational ground state as well as the rotational temperatures of HF were deduced. It was pointed out that the development of the mixing layer along the direction of flow could be divided into three regimes, a laminar region giving rise to a transitional region which would then produce a fully turbulent region. The axial distance for the beginning of transition was estimated to be about 300 boundary layer momentum thicknesses. Regarding the lateral spreading of the mixing layer, it was found to be independent of both the temperature and the velocity ratios of the two streams, although there is some doubt about the spectroscopic measurements in the transition zone.

Grant and also Graham have investigated axisymmetric jets of propane diluted with nitrogen or methane discharging into air at about atmospheric pressure and yielding a lifted diffusion flame. The region of interest was the mixing zone between the exit plane of the jet and the flame location. A variety of instrumentations were used including stroboscopy shadowgraphy, hot wire anemometry, the gated mass-spectrometric technique and Rayleigh scattering. The mass-spectrometric data showed that on the axis of the jet, the air was entrained periodically into the core of the jet and distinct masses of air and fuel persisted without mixing up to the location of the flame. When the mass-spectrometric data were time-averaged, one, of course, observed a great amount of mixing in the time-averaged sense. The formation of ring vortex cores and the presence of unmixed oxygen entrainment were studied when a small velocity fluctuation was impressed on the jet. The structure of the vortex rings in the shear region of the jet flow was observed using Rayleigh scattering. That the vortex rings were stratified was evident from the double peaks in the concentration fluctuation signal. The signal from the jet centerline showed clearly the unmixedness of oxygen and fuel.

The experiments of Libby involving scalar quantities have generally emphasized the significance of intermittency at the outer edge of the mixing layer. One set of experiments dealt with the mixing between helium injected through the wall of a duct conveying air and the other set with the wake of a heated cylinder. The principal measuring tool consisted of a hot wire anemometer probe (a development of the Way-Libby, 1971, probe) which measured two velocity components (u and v) and the concentration at the probe location. In the pipe flow experiments with helium injection, for example, the time series for the variables were obtained and from those the probability density functions (pdf). The pdf can then be employed to obtain the desired moments and cross-correlations. In processing the data, it is implicit that the passage of the *interface* at a given space-point contributes little to the time-averaged value at that point. A significant result from the experiments was that the measured probability density distributions can be very highly skewed in the region of mixing. A near-gaussian and symmetric distribution was observed only where intense mixing had occurred.

III.B. *STRUCTURE OF MIXING LAYERS*

A striking optical observation made in the flow configurations of Brown and Roshko and Winant and Browand is the organization and strength of the large scale structure in the mixing layer. The large scale structure, arising in the absence of any external disturbances, has been shown to persist an appreciable distance and yield a linearly growing mixing layer. The scales and spacings of the eddies do not, of course, increase continuously and eventually, a process of amalgamation of the eddies or a process of *vortex pairing* (Winant and Browand, 1974) sets in, possibly due to the small differences in the eddies.

The presence of orderly structure has also been demonstrated in the mixing experiments involving scalar quantities. In the experiments of Shackleford *et al.*, large scale orderly structure was characteristic of the transition region although the fully turbulent region needs further study. In the experiments of Graham, the presence of an orderly large scale structure was revealed in stroboscopic shadowgraphs. Subsequently, the growth of the interfacial waves at preferred wave lengths and the generation of vortex rings by the folding back of the waves were analyzed by using hot wire anemometry. Over a range of flow rates (or Strouhal numbers based on the diameter of the jet) the waves of the preferred wave length were amplified and one had a nearly single valued spectrum of vortex size. It was pointed out that the interfacial waves and the large scale structure were connected directly, and that the vortex rings formed from the waves were transported over appreciable distances causing entrainment of distinct entities of the surrounding fluid.

Libby in his experiments has taken a different view of the structure of mixing layers. An important possibility in his measurements is the establishment through conditional sampling (Kovasznay, 1970) of the occurrence and non-occurrence of a chosen event at a particular location of the sensor. On this basis, even though it may seem a simplification, it has been suggested that the discriminating techniques yield a type of random telegraph signal for the percentage of time a certain event occurs, for example, the presence of a scalar quantity.

It is too early to draw any quantitative results from the experiments to set up a model for the mixing layer in terms of the behavior of large scale eddies or of the intermittency profiles. If one concentrates on the presence of the large scale structure, it appears that the spacing of the eddies can be related to the mean thickness of the mixing layer in a universal fashion. No such rule has yet been developed for the distance an eddy may be expected to travel before it becomes paired with another. Further experimental studies are required before one can compare the phenomenology of mixing layer turbulence based on incorporating the large eddy mechanics with, say, that based on intermittency. However, a basic point of view is developing that the presence of large scale eddies may be central to the understanding of turbulent mixing layers.

IV. ANALYSIS OF MIXING LAYERS

In general, it is not yet feasible to establish what one wishes to know in turbulent mixing without reference to the governing conditions of specific problems. In fact, there is no adequate generally applicable measure for mixing although as indicated in Chapt. II a start has been made. When scalar quantities and chemical action are present in addition to momentum transfer, the state of complete mixing requires the attainment of homogeneity at the submicroscopic level in the presence of various scales and intensities. Thus, in addition to the time scales of turbulent convection and diffusion, one has to take into account the reactive time scale. One is therefore led to the following hierarchy of problems: 1) determination of momentum transfer in turbulent mixing, 2) determination of scalar fields in the absence of convective motion, 3) determination of the effect of convective motion and 4) modifications introduced by the reaction dynamics. Whether reaction is superimposed on convection and diffusion or the non-linearities are taken into account fully, the ultimate structure of scalar fields depends upon the interplay of the three processes. This can be illustrated in terms of four non-linear interactions.

1) Interaction induced by convection: Suppose we write the conservation equation for a scalar property A in terms of the scalar flux $\vec{q}$ (normal to fluid flow) as follows:

$$(\partial A/\partial t) + \vec{U}\cdot\vec{\nabla}A = D\nabla^2 A \tag{1}$$

$$\partial A/\partial t = -\vec{\nabla}\cdot\vec{q} \tag{2}$$

$$\vec{q} = \vec{U}A - D\vec{\nabla}A \tag{3}$$

It can be seen that the production of scalar fine structure must be important where scalar gradients are small since the action of fluid motion is to produce small scale fluctuations in the A - distribution. On the basis of physical and geometrical considerations, it is postulated then that the perturbations in the scalar distributions are governed by the number and distribution of points in the fluid where the scalar gradient vector is zero.

2) Interaction inducted by the second-order reaction: Suppose we consider a dynamically passive reactant consisting of a single specie undergoing a second order reaction. In a statistically homogeneous scalar field without diffusion, the reaction rate has no dependence on scale and turbulence can only alter the scale of the concentration field. All single point functions of the concentration field (mean concentration, mean square concentration, etc.) will decay (Corrsin, 1958; O'Brien, 1968) at a rate independent of turbulence though not of turbulent convection. Based on an exact result that can be obtained in first-order reactions and assuming that the reactions are not strongly selective spectrally, attempts have been made (O'Brien, 1969) to show that the spectral insensitivity of single species may be quite general.

3) Interaction between the mean concentration and the fluctuations: In the reactive case, it is imperative that one assumes a non-zero mean concentration. Hence, there is always an interaction. It may be observed in regard to the fluctuations that, when the typical amplitude of fluctuations becomes large, a non-gaussian fluctuation distribution becomes unavoidable assuming that the local concentration is always positive.

4) Non-equilibrium effects: Consider a density-sensitive heating mechanism due to chemical action. It can be shown (Eschenroeder, 1965) that in addition to the external source feeding energy into the turbulence, a spectrally-selective internal power source arises. The

external source exerts an intense influence on the energy-containing region of the spectrum and introduces the greatest distortion in a region surrounding the spectral peak. The nonequilibrium effects become dominant as the source effectiveness increases.

We may also note here Corrsin's early decay postulates for the mean concentration and r.m.s. fluctuations, that three asymptotic ranges exist characterized by the ratio of a reaction time scale to a turbulent mixing time scale (see Chapts. I and III for details). The rapid reaction rate is characterized by the fact that the homogeneous kinetic rate becomes quite different from a two-species diffusion-controlled reaction. Since the reaction rate then is controlled by mixing, the depletion rate of the mean concentration has to be related to that of the fluctuations. The resulting local lack of stoichiometry leads to intermittency. Compared to the fast and the slow reaction rate cases, the case where the characteristic times are nearly equal is the most difficult.

IV.A. *MODELING OF CHEMICALLY REACTIVE SHEAR FLOWS*

The simplest and least satisfactory way of calculating such flows is to determine the time rate of production of each species using the instantaneous rate with time mean values of specie concentration, density and temperature. It can be shown (see Chapter III) that the average reaction rate is governed by average concentrations only when the fluctuations do not exist or when they are uncorrelated. In order to take the fluctuations into account, several hierarchies of equations have been developed (O'Brien, 1971; Donaldson and Hilst, 1972; Libby, 1973).

IV.A.1. Higher Order Correlations

Consider a reaction involving two species A and B.

Let the equations governing the time rates of change of the mean values of A and B be

$$d\overline{A}/dt = -k_1(\overline{A}\,\overline{B} + \overline{ab})$$
$$d\overline{B}/dt = -k_2(\overline{A}\,\overline{B} + \overline{ab}) \qquad (4)$$

where A and B are the mass fractions of the species, a and b the fluctuations and k_1 and k_2 the reaction rates. In order to deal with Eq. (4), it is necessary to have a prediction equation for $\overline{ab}$, and it has been shown to be as follows:

$$(d\,\overline{ab}/dt)_{chem} = -k_1(\overline{A}\,b' + \overline{B}\,\overline{ab} + \overline{ab^2}) - k_2(\overline{B}\,a' + \overline{A}\,\overline{ab} + \overline{a^2b}) \qquad (5)$$

which introduces several new terms. While the predicition equations for the r.m.s. variances a' and b' do not introduce any further new terms, one does need prediction equations for the third order correlation terms in Eq. (5).

When can one neglect $\overline{ab}$ in comparison with $\overline{A}\,\overline{B}$? In order to answer this, one can write an equation for the substantive derivative of $\overline{ab}$ in a manner similar to the equation for the Reynolds stress tensor, provided no large gradients are present and assuming incompressible flow for simplicity (Donaldson, 1971):

$$(\partial\overline{ab}/\partial t) + \vec{U}\cdot\vec{\nabla}\overline{ab} = (d\overline{ab}/dt)_{chem} - \text{a diffusion term} \qquad (6)$$

In view of the diffusion term, one can introduce a diffusion length of time as follows:

$$L_D^{\,2} = 2D\tau_D \qquad (7)$$

In a similar manner, considering the first term on the right hand side of Eq. (6), one can introduce a chemical time by writing

$$(d\overline{ab}/dt)_{chem} = -\tau_C(\overline{A}\,\overline{B} + \overline{ab}) \qquad (8)$$

where

$$\tau_C = 1/(k_1\bar{A} + k_2\bar{B}) \tag{9}$$

It is clear that whether the removal of species A and B from the flow is governed by reaction rates or is limited by molecular diffusion is determined by whether the ratio τ_D/τ_C is much smaller or much larger than one respectively. If τ_D/τ_C is very small, $\overline{ab}$ may be neglected and molecular diffusion can be expected to keep the two species well mixed. In assessing a given experiment, one has, of course, to take into account the integral scale of turbulence as well as its relation to the diffusive scale of turbulence.

An alternative way of looking at this question is to examine the reaction rates with and without including the fluctuations. It can be shown (see Chapter III for details) that the concentration fluctuations play little part when A and B are randomly distributed and that the potential for order of magnitude changes in reaction rates arises when the joint distribution of the reactant concentrations are skewed towards large values of these concentrations.

When can the third order correlations be neglected? The third order correlation terms in Eq. (5) control the rate of change of $\overline{ab}$ and hence have a time-integrated effect on it. In order to determine the effect of the higher order correlations, it is necessary to examine the distribution functions from which the mean values and moments are derived (Hilst, 1973). The neglect of the higher order terms can, in general, lead to erroneous results.

The present stage of the Donaldson model for mixing layers is as follows. Separating the diffusion-chemistry model from the turbulence model, one sets up the equation for the rate of change of the velocity-concentration and concentration-temperature terms assuming, say, incompressible flow. One then sets up the equations for the specie concentrations and the variances. Finally, one models the third-order correlations, pressure correlations and dissipation terms in addition to the third-order chemistry correlations and the temperature-chemistry correlation. The set of simultaneous equations is then solved for given initial distributions of the chemical and flow parameters as well as the length scale distributions in the turbulence. This proposal needs further investigation at present.

IV.A.2. Intermittency Model

The hierarchy of equations set up by Libby is for an infinitely fast chemical reaction involving two species and a product under highly diluted conditions. Denoting the mass fractions of the species by A and B, the fast reaction is represented by the relation

$$\overline{AB} \simeq 0 \qquad (10)$$

with the instantaneous production rate of each species given by the combined influence of convection, diffusion and fast chemistry. The equations are unclosed in regard to various order correlations and also the dissipation terms. By introducing closure approximations, the resulting set of equations, involving $\overline{A}$ and $\overline{B}$ and fluctuating quantities in conjunction with the mean element concentrations, can be solved for the mean concentration.

One of the problems with the hierarchy of equations so obtained is the presence of cross-diffusion terms which involve two scalar quantities neither of which is conserved in general. It is conjectured that the cross-diffusion terms arising at the reaction interfaces may be related to the turbulent convection and molecular diffusion at the interfaces, rather similar to the case of passive scalar fields with arbitrary Prandtl number and small scalar gradient. It may be observed that the reaction itself must be confined to the part of the interface where the reactants are contiguous. At the same time, the fast reaction limit requires that the reactants cannot coexist together and the volumes containing one or the other reactant also contain the product. The spreading of product formation thus arises through the straining and molecular diffusion processes in the original volumes.

The infinite reaction rate limit also permits an estimate of a characteristic reaction zone width. This can be done by assuming that the dominant processes within the reaction zone are diffusion and chemical reaction (rather than the convective field) from which the characteristic width of the reaction zone becomes (Pearson, 1963)

$$L_R \simeq \sqrt{D}\ t^{1/6}/k^{1/3} \tag{11}$$

where t is the elapsed time of reaction. On the other hand, if one proceeded according to the uniform strain theory (Batchelor, 1959), the characteristic width of the reaction zone becomes (Gibson and Libby, 1973)

$$L_R \simeq \sqrt{D}\ \gamma^{-1/6}/k^{1/3} \tag{12}$$

where γ is the rate of strain associated with the Kolmogoroff scale eddies. It has been argued (O'Brien, 1973) that the characteristic width should not be related to the uniform rate of strain.

Two other characteristic lengths of significance are $L_B = (D/\gamma)^{1/2}$, the Batchelor diffusion length scale and $L_V = (\nu/D)^{1/2} L_B$, the viscous length scale. Gibson and Libby analyzed the reaction zone formed during the conductimetric titration of a weak base (ammonium hydroxide) and a weak acid (acetic acid) and showed that $L_R < L_B < L_V$.

Another interesting comparison is between L_{NR}, the characteristic length obtained from the time scale for non-reactive mixing and L_R. The rate of loss of a scalar due to simple mixing during an elapsed time period t is $O[tu'/\lambda_s]$ where u' is the r.m.s. turbulence level and λ_s is the Taylor microscale of the scalar field. Comparing that rate of loss with the rate of diffusion of the scalar into a reaction zone, one can show, under certain approximations, that $L_{NR} \simeq L_B N_{Sc}^{-1/4}$ where the Schmidt number $N_{Sc} = \nu/D$. This can be compared with L_R from Eqs. (11) and (12).

IV.B. *MODELING OF TURBULENCE*

The modeling of turbulence is unavoidable in turbulent mixing problems if one wishes to work with transport equations for the Reynolds stress or for the turbulent energy derived by taking moments of the local space-averaged Navier-Stokes equations. There are no intrinsic inter-relationships between the averaged quantities in various moment equations. Thus, in the case of the Reynolds stress balance equation, it is necessary to model the transport and redistribution terms as functions of the Reynolds stresses and the mean velocity field. In addition, we need associated equations for the eddy length scales and the velocity scales. The eddy length scales will, in general, have to be represented by transport equations. Many interesting aspects of such equations can be found in Rotta (1951, 1971), Wolfshtein (1970), Rodi and Spalding (1970), Launder and Spalding (1972), Bradshaw (1972A), Daly and Harlow (1970), Hanjalic and Launder (1972) and Mellor and Herring (1973). One variation in the length scale equation is to employ a variable combining the kinetic energy of turbulence and the length scale.

It may also be mentioned that in developing transport equations, even though the bulk-convection hypothesis (Townsend, 1956; Bradshaw, 1971) has distinct physical appeal, it is the gradient diffusion hypothesis that is the more commonly employed.

In the development of transport equations for scalar quantities, the turbulent diffusion approximation is restricted in application unless the length scale for the scalar quantity and the turbulence energy field variation are small. Nevertheless, an equation for the transport of the mean square fluctuation of a scalar quantity has been obtained (Spalding, 1971) by analogy with the stress equation.

A closure scheme for second order chemical reactions has been proposed (Lin and O'Brien, 1972) in the limit of homogeneous turbulence and isothermal flows, based upon a procedure developed for turbulence dynamics (Orzag, 1971). The closure method has been called the Inequality Preserving Closure Scheme which permits an ordered (though not unique) way of developing closures which satisfy prescribed inequality constraints. The inequalities prescribed are associated with non-negative random variables and the requirement that the closure should asymptotically predict the first and second moments in the limit of molecular diffusion.

IV.C. *SUPERPOSIBILITY OF CHEMICAL ACTION ON TURBULENT MIXING*

It is interesting to examine the question: are there conditions under which ordinary turbulent mixing theory can be combined in a simple fashion with the theory for stochastically distributed reactive flows to analyze reactive turbulent mixing flows (Lin and O'Brien, 1972; Bush and Fendell, 1973; Spalding, 1970; Bilger and Kent, 1972; Libby, 1973)? Some of the difficulties associated with such a postulate should be clear from earlier discussions on nonlinear interactions, the persistence of the influence of higher order fluctuation moments and the non-Gaussian reactant distribution of reactant concentration.

IV.C.1. An Independence Principle

Consider a second-order reaction involving a single specie (A + A → P, *not* A + B → P) in a turbulent convection and molecular diffusion field. The independence principle (Corrsin, 1964; O'Brien, 1969) states that the amplitude of Fourier modes of the concentration in a given flow-reaction configuration is governed approximately by the product of its time history due to mixing and diffusion and its time history due to reaction. This has been discussed in detail in Chapter I. It is unfortunate that there are not enough experimental data to verify the applicability of the hypothesis. The hypothesis seems to have been most successful in the final period turbulence problem when the role of turbulent convection on the scalar spectrum for most wave numbers is less important than the role of molecular diffusion. Some useful information can be obtained on the mean and r.m.s. concentrations through this model, but the role of turbulence and the detailed spectral transfer due to reaction remain unclear.

It is possible to demonstrate the same kind of hypothesis in the case of a stoichiometric reaction involving two reactants and a product also (O'Brien, 1971). In the case of a slow reaction, regular closure schemes seem unavoidable. However, in the case of a fast reaction, one can proceed by distinguishing an initial stage of the reaction from a final stage. It has been shown that the initial stage period (when reaction is the dominant factor in determining the decay of random fields) depends upon the shortest time scale of turbulent convection and diffusion and the reaction rate. Exact stochastic solutions can be shown to exist for appropriate initial conditions. Next, using the solutions of the initial stage as the initial conditions, one can proceed to examine the final stage of the reaction where diffusion is the governing factor. If the diffusivities of the two species are unequal, an approximate solution is all that is possible for mean square concentrations; the time scale of decay of each species will have to be related to the time scale associated with mixing of a single species.

We may note here two special features of the solutions for rapidly reacting mixtures. First, during the initial stage of the reaction, the relative intensity of concentration fluctuations may become quite large depending simply upon the log-normal distribution parameter of the initial state and also it has a lower bound. This fact, combined with the large positive lower bounds for the asymptotic skewness and kurtosis of the concentration fields, seems to indicate that the probability distribution of the concentration is not even approximately normal. Second, when the diffusivities of the two reactants are widely different, the smaller of the diffusivities determines the decay rate of stoichiometric reactions. Thus, single specie experiments may become of value in examining two-specie reactions in this regime.

IV.C.2. Toor's Approach

It will have been realized by now that the basic difficulty in turbulent reactive systems is not the determination of the exchange flux of the species involved, but the establishment of the details of the concentration profiles and the reaction distribution. On a time-averaged basis, turbulent systems differ from non-turbulent systems only in the fact that the interface region in the turbulent system cannot be a thin surface even with rapid reaction. When two turbulent streams come into contact, if the concentrations at the boundaries of the mixing zone are not fluctuating and the diffusivities of the two streams are equal, the mass exchange flux situations for rapid reactions may be calculated in the same manner as in the non-turbulent flow. However, one cannot say that given the time-averaged concentration profiles in a non-reacting turbulent system, one can establish in a corresponding reactive system the time-average rate of reaction even for a rapid reaction. The second approach to this problem - the first being the independence hypothesis- consists in making approximations by means of which a knowledge of the time-average behavior in the equivalent non-reactive system will permit determining the behavior of the reactive system, provided the r.m.s. fluctuations are also known in the equivalent non-reactive system (Toor, 1962). This approach has been discussed in detail in Chapters I and III and reference also may be made to Bush and Fendell (1973).

It may be pointed out that the approximation is a means of obtaining closure of the equations. In applying the independence hypothesis, the first moment equation and spectrum function equation are not closed and some form of closure at a higher moment level is needed. It is given by O'Brien in the form of a consistency condition whereby only the spectral decay due to reaction exhibits wave length similarity with a constant length scale. The approximation introduced by Toor is equivalent to the consistency condition of O'Brien in some respects.

The Toor approach was developed for an irreversible, homogeneous reaction between two highly diluted reactants A and B as follows:

$$A + nB \overset{\overline{k}}{\rightarrow} \text{product} \tag{13}$$

where $\overline{k}$ is the time-averaged reaction rate and n is the stoichiometric ratio. One then considers a non-reactive system which is equivalent to the reactive system in respect to the geometry, turbulent fields and boundary conditions. If we now examine the time-averaged concentration profile in the reactive system, it becomes a function of not only the mean concentration ($\overline{J}$, say, equal to the mean value of $J = nA - B$) but also the relative intensity of the r.m.s. concentration fluctuation. Therefore, it is necessary not only to relate $\overline{J}$ to the time-average behavior of the equivalent non-reactive system, but also to relate the fluctuation intensity of the reactive system to the fluctuation intensity of the non-reactive system. The powerful suggestion of Toor to achieve this was to equate the reactant means to the mean of the reactants. In order to apply this, one has to know $\overline{J}$ and the probability distribution function of concentration which is a function of $\overline{J}$ and the concentration fluctuation. It is therefore of importance to devise experiments by means of which the pdf can be approximated by a suitable modeling of $\overline{J}$. In general, the pdf should be obtained in the hot flame itself. However, it would be useful to determine the extent of the correspondence between the reactive and the cold flow experiments.

There has been an attempt recently (Lin and O'Brien, 1974) to extend the Toor analysis to shear flows. Other than the assumptions of incompressibility, dynamically passive reactants and a one-step, second-order, irreversible exothermic reaction, it is assumed that the heat generated due to chemical action is a constant and that it is the dominant term in the energy balance equation. The temperature fluctuations are taken to be small. Considering a very fast reaction (species segregation and thin interface surfaces) and the non-negativeness of concentration, one can then write

$$\begin{aligned} L(J) &= 0 \\ L(B + P) &= 0 \\ L(B + T) &= 0 \\ L(A) &= W_A \end{aligned} \qquad (14)$$

where P is the product of the reaction between A and B, W_A is the rate of production of specie A in moles, and the operator L is given by

$$L = \partial/\partial t + \vec{U}\cdot\vec{\nabla} - D\nabla^2$$

It may be noted here that the turbulent mixing times are included in the operator L and there is no chemical kinetic scale in the formulation. One can now set up an equation for the pdf in the limit of infinite reaction rate. In order to solve the set of Eqs. (14) and the equation for the pdf, it is proposed that one can set up an analogy between the joint pdf of the coupled reactive system and the joint pdf of a double scalar mixing field, assuming that the velocity fields are identical. In a practical case such as the two-dimensional mixing between two fluids at the same temperature but with different species, the analogous flow becomes one with no species (except the dilutant) with a certain temperature and the experimental concentration in the other fluid. Supposing then the directly measured pdf's of the temperature and the temperature - (non-reacting) concentration fields are available, it is shown that the pdf's of the reactants and the product concentrations can be obtained. Numerical integration of the experimental data will be required in all cases except for the mean concentration of the product.

IV.C.3. Intermittency in Mixing Layers

The final example of superimposing the chemical field on the turbulence field is based on the role of intermittency in mixing layers (Spalding, 1970; Bilger and Kent, 1972; Libby, 1973; Alber, 1973).

Intermittency is a fundamental characteristic of mixing layers, especially when the central mixing layer is bound by interfaces separating the outer non-turbulent flows (Tennekes and Lumley, 1972). The intermittency demonstrates a large scale superstructure upon a fine scale turbulent flow. An important finding in regard to the convoluted turbulent interface is that while intermittency may have no significant effect on the velocity of the fluid in the non-turbulent front, the scalar properties within the same excursions may remain uniform and display a significant change across the turbulent front. If this is finally established, one has to take into acccount the random fluctuations of stoichiometry which are related to the intermittency. Suppose we assume that the instantaneous composition is completely determined by the instantaneous stoichiometry, then it becomes necessary to have some knowledge of the pdf of a parameter indicating the departure from mean stoichiometry. Then, the principal question in including intermittency considerations is establishing the distributions that one can assume for the velocity fluctuations and those that represent the scalar property fluctuations.

Consider a shear flow generated by the mixing of two incompressible, parallel streams with a one-step, very rapid, isothermal chemical reaction involving two highly diluted reactants. The velocity field is first represented by the mean velocity and an assumed distribution of intermittency. The simplest way of doing this is to invoke similarity and obtain the distribution of intermittency obtained in non-reactive shear layers. Thus, let

$$f'(\eta) = f'(0)+(1/2)G[\mathrm{erf}\{\sigma(\eta+\eta_0)\} - \mathrm{erf}(\sigma\eta_0)]$$

$$f = \int_0^\eta f'd\eta \qquad (15)$$

$$\overline{I} = 1 - \mathrm{erf}(\alpha\sigma^2\eta^2)$$

where η is the stretched variable (= y/x), $f(\eta)$ the stream function, G the ratio of the mean velocities of the two streams, σ the spreading parameter of the mixing layer, $\overline{I}$ the intermittency and α a constant of O(1). ( )' denotes differentiation with respect to η.

Considering next the concentration field, the similarity form for the distribution of the mass fractions is written, for example,

$$(f\overline{A})' = (\overline{u_y a}/U_x)' - (R_1 x/U_x) \qquad (16)$$

where U_x is the velocity of the reference stream and R_1 is the mean rate of product generation. It is then argued that if we know $\overline{A}$ by whatever means, then the mean specie generation $(R_1 x/U_x)$ can be found as a function of η.

Suppose now we introduce a conserved scalar quantity ξ defined by

$$\xi = B - (W_B/W_A)A$$

where W_A and W_B are the molecular weights. It follows that at a particular space-time point, $\xi > 0$ implies $\xi = B$ and $\xi < 0$ implies $\xi = A$. In other words, at a given space-time point, $\xi > 0$ indicates passage of eddies with reactant B and product and $\xi < 0$ indicates passage of eddies with reactant A and product. The mean concentrations can therefore be related to the conditional probability

$$\bar{B} = E(\xi | \xi > 0)$$

$$\bar{A} = (W_B/W_A)E(\xi | \xi < 0)$$

where E is the expected value of ξ. The conditional events are given to the right of the vertical bar.

In addition to ξ, one can define the variables

$$Z_A = A + (W_A/W_P)P$$

$$Z_B = B + (W_B/W_P)P$$

where W_P is the molecular weight of the product P generated from chemical action between A and B. The conservation equation for Z_i (i = A, B or P) is

$$f\bar{Z}_i' = (\overline{u_y Z_i}'/U_x)' \qquad (17)$$

The entire concentration field may therefore be found from a knowledge of Z_i and Eqs. (16) and (17).

The fundamental question then is how to establish by theory or experiment adequate information regarding the pdf of ξ, the mean values of Z_i and ξ and the first few moments of the fluctuations. Several models for the pdf of ξ have been proposed incorporating the intermittency of the flow. However, there is not yet a convincing means of accounting for the intensity of the scalar variable fluctuations. Even the mean values of Z_i and ξ can be established only on heuristic grounds.

IV.D. *COMPRESSIBILITY EFFECTS*

Brown and Roshko (1971) conducted experiments in two interesting limiting cases: 1) the low speed isothermal mixing of two gases of different molecular weights and 2) the compressible adiabatic mixing of a high speed flow with the same gas in a quiescent state. The experimental results seem to indicate little alteration in the spreading rate with large density differences in the first case and a significant reduction in the spreading rate with Mach number in the second case. Experimental evidence is also available that a light gas spreads relatively quickly into a heavier gas (Chriss and Paulk, 1972) and a heavy jet spreads relatively slowly (Abramovich, 1969). This subject has remained controversial over the years. In spite of the great interest in high speed, large enthalpy reacting flows, there have been few measurements of a detailed nature.

Compressibility effects have generally been accounted for on the basis of the Morkovin (1962) hypothesis based on the smallness of pressure fluctuations at small Mach numbers. When streamwise density gradients are present, it appears that one can still employ Morkovin's rule, though there is experimental evidence (see Bradshaw, 1972B) that mean dilatation does affect the structure of turbulence. It has therefore been argued that the effect of dilatation can be taken into account by increasing the dissipation term in a semi-empirical manner.

IV.D.1. Exact Equations

In examining mean compressible turbulent flows, the turbulent stress tensor can be shown (Laufer, 1969) to take the same form as the incompressible stress tensor, and furthermore, the addition of mean turbulent stresses acting on fluid elements along mean streamlines can be shown to be consistent both with respect to momentum balance and mechanical energy balance. However, we do not have adequate information concerning the importance of turbulent dissipation and pressure diffusion terms in relation to the production terms.

Bray (1974) has deduced the exact two-dimensional equations for a turbulent reacting flow. The exact equation of Bray for the turbulence energy balance, for example, deals with a balance among 1) the time rate of change of mean kinetic energy per unit volume, 2) the convective rate of change due to mean velocities, 3) the convective rate of change due to turbulent mass transport fluxes, 4) the production of turbulence energy due to buoyancy effects, 5) the production of turbulence energy due to work done by turbulent stresses in velocity gradients utilizing the kinetic energy of the mean motion, 6) the viscous diffusion, 7) the viscous dissipation of turbulence energy due to heat, 8) the transfer of energy from the enthalpy of the gas to turbulence, 9) the non-stationary transfer of energy from turbulence to mean motion, 10) the non-stationary gain of turbulent kinetic energy per unit volume of gas and 11) the turbulent diffusion of turbulence kinetic energy. An order of magnitude analysis of that equation has been carried out based on the following assumptions: 1) the shear layer is governed by boundary layer approximations, 2) the characteristic fluctuation magnitudes may be related to the ratio of the shear layer width to a characteristic length, 3) two turbulence length scales can be identified which are related to the turbulence kinetic energy balance and 4) a distributed time-average chemical reaction zone exists. One can then apply the usual closure hypotheses and obtain equations in more familiar forms in incompressible flow (e.g. Harsha, 1973). However, the order of magnitude analysis still applies to low Mach number flows.

V. STATISTICAL CONTINUUM THEORIES

In a broad sense, the problem of nonhomogeneous turbulent mixing can be formulated in terms of two questions: 1) given a body of fluid in contact with another (in each of which motions can be expected to conform to certain equations) and given that, at some initial instant, the velocity of the fluid is a random function of position described by certain probability laws, to determine the probability laws that govern the motion of the fluid in the contact region at subsequent times; and 2) given a statistical description of each of the species present in the two bodies of fluids and given the instantaneous rates of reaction in terms of kinetic equations, to determine the probability laws that describe the distribution of each of the species at subsequent times. We are concerned here with the second question.

It is the general objective of statistical continuum theories to obtain a set of determinate equations for turbulent flow systems in contrast to the Reynolds transport equations which contain undetermined quantities. However, in view of the dynamic nonlinearities (with or without a chemical reaction) in turbulent flow, the closure difficulties of the moment equations are formally the same as in the moment equations deduced from the Navier-Stokes equations for the turbulence problem. We shall discuss three statistical approaches, two of them based upon a distribution function approach and the third based on a model equation for the distribution function of the velocities in turbulent flow.

V.A. *EDWARDS' MODEL*

The model is based on the premise that the problem of turbulence can be cast in the form of differential equations in function space using an appropriate probability density functional (Hopf, 1952; Chandrasekhar, 1943; Kraichnan, 1959; Edwards, 1964). In particular, one can consider the problem of a randomly excited turbulence, the forcing function being related in shear flow to the mean rates of strain. Starting with a Liouville-type equation, itself deduced from the Navier-Stokes equations in function space, it has been shown that 1) the resulting nonlinear integral equation for the energy spectrum of the velocity field is analogous to the Peierls-Boltzmann equation for phonon scattering; 2) the Boltzmann-type equation may be replaced by a local differential equation, namely, the Fokker-Planck equation; and 3) an analogue of the Fokker-Planck equation can be derived for the energy occupation of wave-number space, noting that the energy cascade is local in wave number.

It is then found (Edwards and McComb, 1971) that the notions of a generalized diffusivity and a generalized viscosity are fruitful and mathematically tractable in the steady state; the introduction of a generalized viscosity is the principal artifice in obtaining closure of the equations. This formulation still leaves completely open the question of a connection between the generalized diffusivity and viscosity. It has been argued that the generalized viscosity may be selected on the basis of maximizing the entropy of turbulence, though this is open to question in some respects.

Finally, it has been shown (Edwards and McComb, 1972) how differential equations (in centroid variables) can be obtained for the mean velocity, kinetic energy and shear stress by the expansions of the nonlinear integral transport equations in a series of homogeneous kernels.

V.A.1. Hill's Extension

An extension of this approach to chemically reactive systems has been attempted (Hill, 1969). It is assumed that the molecular diffusivity of the species and the reaction rate are constant. However, the invariance properties and the inequalities have to be satisfied; in particular, all statistical functions of the concentration have to be independent of the velocity field in homogeneous turbulence in the limit of zero diffusivity (O'Brien, 1966 and Chapter I).

The mean decay rates and the two point correlation functions of the concentration field can be obtained from the probability density functions. In order to determine the one and two-point pdf's of the concentration field, one can set up the Liouville equation, as discussed earlier, for the conditional probability functional. The single point pdf is then obtained by multiplying the Liouville equation by the full pdf of the velocity field for all time. The resulting equation includes the convective and the diffusive terms. The convective term has been approximated by an expansion as stated earlier. The diffusion term is more difficult to handle. In any case, once a hierarchy of multi-point pdf equations is obtained in x-space, any closure of the hierarchy can be shown to preserve the zero diffusivity invariance and to yield the exact result in the absence of diffusion. Further, in the case of a second order reaction, if the independent variable, namely, concentration, is replaced by the specific volume of reactant, the moment equations satisfy the invariance, are closed at each level and give the exact result when diffusion is neglected.

V.B. *O'BRIEN'S MODEL*

The second approach (Dopazo and O'Brien, 1974) is also based upon a functional equation formulation (see Chapter I). A one-step, second-order, irreversible, exothermic chemical reaction (dynamically passive reactants) is examined using the pdf of the temperature and the concentration of species. The very rapid reaction examined is for the ignition period of the mixture. The velocity field is assumed to have zero mean and to be unaffected by mass production or chemical heat. The Arrhenius rate constant is taken to be temperature dependent. The molecular diffusivities are assumed equal and the Lewis number is taken as unity. An equation for the single point pdf is then generated from the conservation equations, in a way rather similar to the approach of Edwards, under the assumptions of 1) negligible reactant consumption, 2) homogeneity and 3) the smallness of the correlation between the temperature and the concentration fields initially and during the initiation of the fast reaction.

The equation for the pdf is again not closed on account of the presence of the two point joint pdf in the diffusion term. It may be pointed out that if convection had been included, a second term (the nonlinear convective term) would have arisen in the equation and so there would have arisen a second cause for the equation to remain unclosed.

The closure for the equation with only diffusion is obtained by assuming a Gaussian conditional expected value for the temperature at a point, given the temperature at a neighboring point. This method of obtaining closure is similar to the one employed by Lundgren (1969) who introduced the BGK relaxation term.

Finally, the probability density function equation may be written as follows:

$$\partial P(T,t)/\partial t + C_1\partial[e^T P(T,t)]/\partial t = C_2\partial\{[T - \overline{T}(t)]P(T,t)\} \quad (18)$$

where $P(T,t)$ is the pdf of the temperature field at time t, C_1 is a constant independent of time and C_2 is the inverse of the second Damköhler number based on the dimensional Taylor's microscale, the latter taken to be a constant.

The introduction of the dimensional microscale into the second Damköhler number is based on the suggestions of

Corrsin (1958) and Kovasznay (1958). First, the microscales of the scalar and the velocity fields are related through the Schmidt number. Second, it is suggested that the intensity of turbulence in itself is insufficient to characterize the details of the turbulence and that the velocity gradient or the vorticity based on the microscale is also significant. However, there is no critical test to distinguish the influence of the turbulence scale and the microscale.

The first-order hyperbolic differential equation (18) with variable coefficients can then be investigated to determine the evolution of P(T,t) with time. However, the actual computation has presented several difficulties.

The application of the foregoing approximate method to turbulent shear flows is yet to be demonstrated.

V.C. *CHUNG'S MODEL*

The third approach of interest here is one (Chung, 1970, 1972, 1973) wherein a model equation is developed for the distribution function of the velocities in turbulent flows.

The fundamental bases of this approach are the following: 1) At high Reynolds numbers, a statistical separation exists between higher equilibrium wave numbers and lower non-equilibrium wave numbers and the generalized Brownian motion is an adequate description of the result (in the spectral plane) that the non-equilibrium degree of freedom decays through its interaction with all other degrees of freedom. It is an acceptable description once the clear separation in the characteristic times of the large and small eddies is recognized. 2) The transport of various properties in turbulence is due to the larger eddies and one can assume that the statistical property of a fluid element is substantially due to the larger eddies. It is, therefore, possible to consider a single significant dynamical scale associated with the larger eddies. If one postulated more than one dynamical scale, it is necessary to establish some connection between the two scales (Bywater and Chung, 1973). There is at present no systematic way of relating such scales. 3) In view of 1) and 2), one can (see Chandrasekhar, 1943) set up Langevin's linear stochastic equation to represent the decay of the (fluid element) momentum associated with the lower nonequilibrium wave numbers. The important boundary condition for this equation is that as the characteristic decay rate of lower wave numbers tends to infinity and the molecular dissipation tends to zero, the distribution function for a transferable property must become Maxwellian. It will be noted that the distribution function is introduced as a weighting function and in an average sense. In the final analysis, we are concerned only with certain moments of the distribution function. 4) The distribution function can then be described by a modified form of Fokker-Planck equation which is equivalent to the Langevin's equation. That integral equation is converted into a differential equation by Taylor series expansion. 5) The various order moment equations (deduced from the generalized moment equation corresponding to the differential equation) can be shown to be not totally inconsistent with the corresponding order moment equations deduced from the Navier-Stokes equations. In fact, the characteristic times for the equilibrium and nonequilibrium degrees of freedom, the molecular dissipation and the molecular diffusion are obtained from such consistency considerations and assumptions pertaining to the characteristic lengths of the larger eddies and the applicability of the universal equilibrium theory.

One of the important steps in the foregoing is the

introduction of the Fokker-Planck equation for the distribution function. One argument for the introduction of that equation is the assumption that pressure fluctuations have a randomizing effect. In other words, the Fokker-Planck equation is a means of modeling eddy scrambling. In applying this theory to chemically reactive flows, the distribution function of the transferable quantity is written as follows:

$$F^{(\alpha)}(X_i, U_i, t) = n^{\alpha}(X_i, U_i, t) f(X_i, U_i, t) \qquad (19)$$

where f $\underline{dx}$ $\underline{du}$ is the occupancy probability of the fluid element, n the concentration of a transferable property and α is a constant equal to or greater than zero. The probability average of n^{α} is then

$$\langle n^{\alpha} \rangle = \int F^{(\alpha)} \underline{du}. \qquad (20)$$

Now, one can deduce the generalized moment equation including n and compare the different order moment equations obtained therefrom with the moment equations obtained from the classical equations.

In the case of a single chemical specie diluted in an inert medium, the modified Fokker-Planck equation for f is complete in itself for the so-called homologous case -- stationary, homogeneous field with uniform velocity and concentration gradients. The determination of the concentration field is then straight-forward in problem formulation. The next order of complication arises when we consider, for example, two uniform, isotropic, parallel streams in contact which yield a homologous region both with respect to the velocity and the concentration fields. The two parallel streams could have different but at the same time constant velocities and concentrations. If the concentration in the two steams is totally independent of the velocities therein, no inhomogeneities will arise in the mixing layer.

Finally, one has the case where chemical reaction can occur between the two species. Any formulation of that problem within this framework will require assumptions pertaining to 1) the Damköhler number, 2) the molecular Schmidt number, 3) the mean gradients of velocity and concentration in the mixing layer and 4) the coupling between the concentration fluctuations and the structure of turbulence in the total flow field of interest. Hence, the turbulence transport and the mean square fluctuation of concentration will become a function of those parameters.

V.D. *SPECTRAL STRUCTURE OF SHEAR FLOWS*

In view of the importance of the behavior of wave number spectra in the basic studies on turbulence, we shall touch briefly upon the spectral structure of turbulent scalar and reactive fields. It is well known (Kolmogorov, 1941) that at high Reynolds numbers, the small scale components of turbulence (small eddies, high wave numbers) can be considered to be steady, isotropic and independent of the large scale components (large eddies, low wave numbers) and that energy is transferred from the larger eddies to the smaller eddies. The behavior of the small scale eddies can be obtained explicitly in the inertial subrange where dissipation is negligible and the transfer of energy is dominantly by inertia forces. The dissipation range is much more difficult to handle. The theories (for example, Obukhov, 1941; Heisenberg, 1949) developed for the universal range are unlikely to be applicable in the dissipative range. Various phenomenological (Townsend, 1951) and approximate theories (Kraichnan, 1959, 1962; Edwards, 1964) are available.

In respect to scalar quantities such as concentration and temperature, the inertial subrange has been investigated (Obukhov, 1949; Corrsin, 1951) to determine the small scale structure of the fluctuating field. The spectra for scalar quantities has been studied (Corrsin, 1964) for turbulent mixing with second-order chemical reaction in the universal range. Reference should be made to Chapter II for details regarding these.

Another approach to the study of the transfer of turbulent kinetic energy and scalar quantities at large wave numbers is based upon a simple continuous spectral cascading process (Pao, 1965). The entire universal equilibrium range of wave numbers (k) can be covered by this. Considering that the cascading is mainly due to turbulent convection which in turn is dependent upon ε, the cascading rates are assumed to be dependent upon ε and k.

For a second-order chemical reaction involving reactants and product, one can then obtain the spectrum function as follows: 1) In the universal equilibrium range, one considers $k > k_s$, where k_s is based on the smaller of the integral scales associated with the scalars. One finds then

$$F_s(k) \propto k^{-7/3}\exp\{-(3/2)A_1k^{4/3} + 3A_2k^{-2/3}\} \qquad (21)$$

2) In the inertial convective range, with diffusivity and viscosity negligible,

$$F_s(k) \propto k^{-5/3}\exp(3A_3k^{-2/3}) \qquad (22)$$

and 3) In the viscous-convective and viscous-diffusive ranges, if we assume that the cascading rate depends upon the local least turbulent straining rate,

$$F_s(k) \propto k^{n_1}\exp(-n_2k^2) \qquad (23)$$

where n_1 is related to the diffusivity and the strain rate and n_2 is inversely proportional to the strain rate.

VI. TURBULENT COMBUSTION

It is convenient to divide combustion systems into premixed systems and nonpremixed systems even when finite rate chemistry is considered. It is necessary to distinguish between combustion in laminar flows, turbulent flows and transitional flows from laminar, turbulent or transitional combustion in given flows. However, if the reactants and products are assumed to be dynamically passive, we can restrict attention to the influence of the flow field on the reaction. We have very little understanding of the influence of combustion on turbulence, transition or laminarization. That there should be mutual interaction between the chemical and the flow fields is obvious. Bray (1974), on the basis of experimental observations made by Scurlock and Grover (1953), Westenberg (1954, 1959), Gunther and Lenze (1973) and Eickhoff (1973), points out that 1) the shear produced within a flame will generate turbulence and 2) turbulence energy will be removed due to velocity divergence (resulting from heat release), diffusion (arising from density gradients set up within the flame) and viscous dissipation (due to reduction in turbulence Reynolds number). However, these are general observations and one is not yet able to obtain a unified picture of premixed and diffusion flames or open and confined high-speed flames.

VI.A. *LAMINAR AND TURBULENT COMBUSTION*

It is considered useful in several respects to recall certain features of laminar combustion as a basis for examining combustion in turbulent flows. The general method of attack for combustion in laminar flows seems to be well established (Williams, 1971). It is possible to take into account the nonisothermal reaction kinetics, variable properties and arbitrary orders of reaction within the limitations of regular and singular perturbation techniques and similarity assumptions.

In comparing laminar and turbulent flames, the notion of a turbulent flame speed is introduced by several investigators. It is deduced from the mean motion of the visible flame boundary and is generally thought to be related to the turbulence intensity. However, experimental evidence exists (e.g. Ballal and Lefebre, 1973) which indicates that the so-called turbulent flame speed varies in a complex manner both with the intensity and the scale of grid-generated turbulence. It appears, therefore, that turbulence flame speed is meaningful only when applied to the same class of configurations.

It has been pointed out by Williams (1974) that in setting up a species conservation equation, for example, in a reaction involving species A and B giving rise to a product P, that it may become necessary to take into account 1) a convective term, λ_1, for example, in a premixed flame and 2) shear or flame stretch depending on the flow conditions. Regarding the latter, laminar flames can be shown (Bush and Fendell, 1970) to consist of two zones, an upstream convective-diffusive zone (with negligible reaction) and a reactive-diffusive zone (where convection may not be significant). In the convective-diffusive zone, the shear determines the evolution of the flame. For positive flame stretch, convective-diffusive balance can be established on both sides of the reaction zone in the diffusion flame.

A point of some importance here is that large positive strain rates will tend to cause extinguishment. In a diffusion flame, a stretched condition causes greater reactant flux into the reactant zone; this decreases the residence time therein compared to the chemical reaction time and, hence, there is insufficient time for heat release. If γ is the flame stretch indicating the rate of increase of the flame surface area and τ_{ch} is the chemical reaction time, it is significant to consider when $\gamma\tau_{ch}$ < the Damköhler number for extinction. In the case of two steams of reactants which are coming into contact in a two-dimensional mixing layer (Linan, 1963; Phillips, 1965; Clarke, 1967), one can calculate (Williams, 1971) the formation of the laminar premixed flame zone, its extension to the diffusion layer, its extinguishment and the downstream progress of the diffusion flame.

The structure of combustion zones in turbulent flow can be understood in many respects by examining the turbulence length scales and the strain rate. For example, the applicability of the wrinkled laminar flame model may be related to the flame thickness being less than the Kolmogoroff length scale. Flame stretch is a useful concept both in determining the smoothing and thickening of the flame profiles (negative strain rate) and the extinguishment of hot spots (large positive strain rates). For example, one can consider (Bush and Fendell, 1973; Libby, 1972) that in a mixing layer formed in a splitter plate experiment (Brown and Roshko, 1971), for Prandtl and Lewis numbers equal to unity, the reactants coexist only in the vicinity of the vortex sheet (on an instantaneous basis); and under a fast reaction approximation, one can identify the behavior of the diffusion flame in analogy with the behavior of the interface. Thus, the diffusion flame structure may be governed largely by the local strain rate, and this stretching of the flame leads to a lengthening of the flame, an increase in reactant consumption and, when the strain rate is too large, an annihilation of the hot spot. The behavior of a diffusion flame under a finite tangential strain in an unbounded counter flow has been examined (Carrier *et al.*, 1973) using a singular perturbation analysis; such an analysis permits a reaction zone of small but finite thickness.

VI.B. *MODELING OF TURBULENT FLAMES*

VI.B.1. Spalding's Model

One scheme that has been evolved for turbulent diffusion flames (Spalding, 1970) consists in predicting the complete flame properties, under the assumption of very fast chemistry, by solving six simultaneous equations for six variables as follows: three for the hydrodynamic field (time mean value of velocity, turbulence energy and length parameter) and three for reaction (stagnation enthalpy, mixture fraction and concentration fluctuation squared). The mixture fraction represents the fraction of all of the material in a sample which has emanated from one of the supply sources; it is not the same as the mass fraction of either stream. One of the features of the model is that it yields a diffusion flame of finite thickness, a consequence of the large fluctuations of concentration. At any section along the flow, there is a radial distribution of temperatures, the maximum, the mean and the minimum; that is, at any section along the flow, in any radial position, the mixture ratio is stoichiometric for a finite fraction of time.

Spalding (1971) has also suggested that the major effect of turbulence on a chemical reaction can be understood on the basis of an "eddy break-up" model. It is assumed that 1) the rate of consumption of fuel is related to the rate at which the fuel concentration fluctuations disappear, 2) an analogy can be set up between energy transfer during eddy break-up and the rate at which unburned gas lumps are broken down into smaller sizes and 3) the fuel concentration fluctuations can be calculated from the mass fraction of fuel. The rate of energy decay, when the local generation rate of turbulence equals the local decay rate, is approximately equal to $0.35\rho|\partial U/\partial y|$ where $|\partial U/\partial y|$ is the local velocity gradient. Defining a reactedness R* by

$$R^* = (A - A_u)/(A_b - A_u) \tag{24}$$

where A is the mass fraction of fuel and b and u represent the burned and unburned parts, the time-mean volumetric rate of fuel consumption can be written as

$$\{(R^*\dot{A}_{max})^{-1} + [C(1 - R^*)\rho|\partial U/\partial y|]^{-1}\}^{-1}$$

where A_{max} is the maximum rate of fuel consumption by chemical action and C is a constant of the order of 0.35. The implication here is that the volumetric rate is the sum of that due to chemical kinetic processes and eddy break-up processes. It will be observed that A(= dA/dt) is zero for the gas both with R* equal to zero and one. At some interface between lumps of gas, the reactedness may be, it is presumed, such that (dA/dt) has a maximum value. The reactedness R* itself is described as a measure of the fraction of time at which high reactedness gas is present.

The model was developed originally for confined premixed turbulent flames (e.g. experiments of Howe *et al.*, 1963 and Cushing, *et al.*, 1967). Some suggestions for improving this model have been made. One is to set up some connection between the turbulence and the reaction rate, perhaps through a local Reynolds number. Another is to associate reactedness with the gas locally and set up a transport equation for the root mean square fluctuation of the reactedness.

VI.B.2. Boundary Layer Formulation for a Fuel-Oxidizer Configuration

A model for the interaction of turbulence and chemical reactions has been developed by Rhodes and Harsha (1972). It is a probabilistic model based on heuristic arguments. Several interesting features are also incorporated into the closure of the momentum equation (see Rhodes *et al.*, 1974), but we shall discuss here the chemistry model.

In the usual formulation of a chemically reactive problem, one sets up the specie continuity equation with an appropriate production term. One can replace that by means of an element concentration equation which also represents the mean total enthalpy if the Prandtl and Lewis numbers are taken to be unity. However, the average density at a point cannot be related directly to the average total enthalpy and the average element concentration. The central question then is how to specify the average density with turbulent equilibrium or frozen chemistry and with turbulent finite rate chemistry. In essence, the average density is determined not from the average of species and enthalpy at the point under consideration, but from an average taken over the species and enthalpy values which contribute to the averages at the point. As a consequence of this procedure, one can model the diffusion and production of species without the necessity of including the species continuity equation; it is not necessary to define profiles for the species.

The mixing layer is divided simultaneously into a discrete number of zones and classes. A zone is an elemental physical region characterized by a time-averaged elemental composition. The distribution of the zones is determined by the spatial distribution of stoichiometry. A class is fluid located anywhere, which can be characterized by an instantaneous elemental composition.

When there is mass addition to a class, the added mass is of the same elemental composition as the class, by definition. The material of the class may be distributed throughout the mixing layer but is most abundant in the zone where the average stoichiometry is the same as the class stoichiometry. The majority of a class may or may not be formed in that zone.

It is assumed in the model that a single density may be assigned to a fluid in a class. The velocity assigned to a class is that of the zone with the same elemental composition as the class even though the correlation between species and velocity fluctuations is not unity. The class total enthalpy is assumed to be linearly related to the class elemental composition and this assumption is consistent with Lewis number being unity and the total enthalpy being large compared to the energy dissipation.

There is a finite probability, P_{ij}, in turbulent flow of finding class j in any zone i. If the pdf of concentration is P_{Ai}, then

$$P_{ij} = \int_{A_j - \Delta A/2}^{A_j + \Delta A/2} \tag{25}$$

where A_j is the instantaneous elemental composition. The time-averaged values may be defined by

$$\overline{A}_i = \sum_j P_{ij} A_j \tag{26}$$

$$\overline{\rho}_i = \sum_j P_{ij} \rho_j \tag{27}$$

In order to use Eq. (26), it is necessary to make some approximation for P_{Ai}. In view of the lack of necessary data, this is done rather arbitrarily at the moment, for example, by relating P_{Ai} to the standard deviation and relating the standard deviation for concentration to that for the fluctuating velocity. One can then determine the time-averaged density in an equilibrium or frozen flow.

In order to extend this approximation to a situation with finite rate chemistry, each class is considered as a transient perfectly stirred reactor. However, this development is still in its initial stages.

VI.B.3. Intermittency Model

Another model for turbulent flow combustion is based (Libby, 1973, 1974) on the concept of the "oscillation" of a flame sheet within the mixing zone formed between two parallel streams each containing one of the reactants. There is emphasis here on the intermittent nature of the turbulent flow and, hence, the structure of the turbulent-non-turbulent interfaces and also of the flame sheet. We shall note at the outset that the two reactants are assumed to be involved in a one-step, single-product, fast reaction.

It is significant to note here that the reaction surface is assumed to be a convoluted oscillating surface, the dynamical nature of which is determined by the presence of the turbulent strain and mechanism associated with it. However, the convoluted oscillating reaction surface or the outer interfaces of the mixing zone are *not* viewed as involving the large scale structure that has been observed (Brown and Roshko, 1971; Winant and Browand, 1974) in the development of the mixing layer. A distinction is made between the possibility of the existence of contiguous large eddies (one with one reactant and an adjacent eddy with another reactant, with dilutent and product in each) and the possibility of an oscillating reaction surface. Whether there is a connection between the mechanics of the large scale structure and the mechanics of the strained convoluted surface is not yet clear. The model is based on interpreting the oscillation of the reaction surface in terms of the pdf of the mass fractions of the two reactants, which, in turn, is related to the intermittency of the flow. The pdf is *not* assumed to be Gaussian or near-Gaussian. Suppose we have the concentration time-history at a point in the mixing layer, one can develop a zero-one discriminating function to identify the species and then obtain zone averages, crossing frequencies and point statistics by the usual methods of conditional sampling. One can also deduce the pdf based on an adequate number of traces.

In developing an analysis based on this model, one proceeds, as discussed earlier, by establishing under appropriate conditions a connection between the behavior of chemically passive scalars and chemical reactants. Next, one has the problem of estimating the pdf of a concentration variable. Information pertaining to the mean values and fluctuation intensities (taking into account intermittency) of the concentration variables then becomes central to the modeling of the pdf.

Another difficulty should also be pointed out in the case of diffusion flames (Williams, 1974). One can obtain the heat release per unit area of a flame sheet by making it proportional to $D_A \equiv \rho D|\nabla A|$, which is evaluated at the stoichiometric value; ρ and D are the local density and diffusion coefficient, respectively. The distribution function for the quantity D_A can be obtained from the joint distribution function for A and $|\nabla A|$ which itself can be replaced by the conditioned density at A equal to the stoichiometric value. In experimentation, this calls for advances both in sensing and in data processing. However, the problem is still to relate the energy release per unit area to the volumetric energy release. In order to do that, it is essential to take account of the non-Gaussian character of the distribution function which is evident in measurements.

VII. CONCLUSION

The modeling of turbulent mixing has progressed beyond the possibilities of the 1968 Stanford Conference. Many problems of practical importance can now be modeled for numerical solution. The past ten years have brought enormous advances in computers and computational techniques on the one hand, and in measurement and data processing on the other. However, the mutual interaction of turbulence and chemistry is an extremely difficult problem. In general, where scalar quantities are involved, while several models have been developed for mixing based on heuristic grounds, extensive measurements are required before the relation between entrainment, molecular mixing and the rate of chemical action can be fully understood.

VIII. REFERENCES

Abramovich, G.N. (Editor) (1969) TURBULENT JETS OF AIR, PLASMA AND REAL GASES, Consultants Bureay, New York.

Abramovich, G.N. (1963) THE THEORY OF TURBULENT JETS, M.I.T. Press, Cambridge, Massachusetts.

Alber, I.E. (1973) TRW Systems Report 188117-6019-RU-00, Redondo Beach, Calif.

Ballal, D.R. and Lefebre, A.H. (1973) *Acta Astronautica 1,* March-April 1974, 427.

Batchelor, G.K. (1959) *J. Fluid Mech. 15,* 113.

Bilger, R.W. and Kent, J.H. (1972) Charles Kolling Research Lab. Tech. Note F-46, The University of Sydney, Sydney, Australia.

Birch, S.F. and Eggers, J.M. (1972) NASA AP-321, 11.

Bradshaw, P. (1971) AN INTRODUCTION TO TURBULENCE AND ITS MEASUREMENT, Pergamon Press, London.

Bradshaw, P. (1972A) *Aero. J. 78,* 403.

Bradshaw, P. (1972B) I.C. Aero Report 72-21, Imperial College of Science and Technology, London. See also (1974) *J. Fluid Mech. 63,* 449.

Bray, K.N.C. (1974) AASw Report No. 332, University of Southampton, Department of Aeronautics and Astronautics.

Brown, G. and Roshko, A. (1971) AGARD CONFERENCE PROCEEDINGS No. 93, London, 23-1.

Brown, G.L. and Rebello, M.R. (1972) *AIAA J. 10,* 649.

Brown, G.L. and Roshko, A. (1974) *J. Fluid Mech. 64,* 775.

Bush, W.B. and Fendell, F.E. (1970) *Comb. Sci. Tech. 1,* 407.

Bush, W.B. and Fendell, F.E. (1973) *Phys. Fluids 16.*

Bywater, R.J. and Chung, P.M. (1973) AIAA Preprint Paper No. 73-646.

Carrier, F.G., Fendell, F.E. and Marble, F.E. (1973) Project SQUID Tech. Rep. TRW-5-PU, Project SQUID Headquarters, Purdue University, West Lafayette, Indiana.

Chandrasekhar, S. (1943) *Revs. of Mod. Phys. 15,* 1.

Chriss, D.E. and Paulk, R.A. (1972) AEDC-TR-71-236, Arnold Engineering Development Center, Tullahoma, Tennessee.

Chung, P.M. (1970) *Phys. Fluids 13,* 1153.

Chung, P.M. (1972) *Phys. Fluids 15*, 1735.

Chung, P.M. (1973) *Phys. Fluids 16*, 1646.

Clarke, J.F. (1967) *Proc. Roy. Soc. London, A296*, 519.

Corrsin, S. (1951) *J. App. Phys. 22*, 469.

Corrsin, S. (1958) *Phys. Fluids 1*, 42.

Corrsin, S. (1964) *Phys. Fluids 7*, 1156.

Cusing, B.S.,Faucher, J.E., Gandbhir, S. and Shipman, C.W. (1967) ELEVENTH SYMPOSIUM ON COMBUSTION, Combustion Institute,817.

Daly, B.J. and Harlow, F.H. (1970) *Phys. Fluids 13*, 2634.

Donaldson, C.duP. (1971) AGARD CONFERENCE PROCEEDINGS No. 93, London, B-1.

Donaldson, C.duP. and Hilst, G.R. (1972) PROC. OF THE 1972 HEAT TRANSFER AND FLUID MECHANICS INSTITUTE, Standford University Press, Stanford, Calif., 256.

Donaldson, C.duP. (1973) in AMERICAN METEOROLOGICAL SOCIETY WORKSHOP IN MICROMETEOROLOGY, Edited by D. A. Hangen, Science Press, New York, 313.

Dopazo, C. and O'Brien, E.E. (1974) *Acta Astronautica 1*, 1239.

Edwards, S. (1964) *J. Fluid Mech. 18*, 239.

Edwards, S. and McComb, W.D. (1971) *Proc. Roy. Soc. London A325*, 313.

Edwards, S.R. and McComb, W.D. (1972) *Proc. Roy. Soc. London A330*, 495.

Eickhoff, H.E. (1973) COMBUSTION INSTITUTE EUROPEAN SYMPOSIUM (Ed. F. J. Weinberg), Academic Press, New York, 513.

Eschenroeder, A.Q. (1965) *AIAA J. 3*, 1839.

Gibson, C.H. and Libby, P.A. (1973) *Comb. Sci. Tech. 6*, 29.

Graham, S.C., Grant, A.J. and Jones, J.M. (1975) Submitted for publication in *AIAA J*.

Grant, A.J., Jones, J.M. and Rosenfield, J.L.J. (1973) COMBUSTION INSTITUTE EUROPEAN SYMPOSIUM, Academic Press, London.

Gunther, R. and Lenze, B. (1973) FOURTEENTH SYMPOSIUM ON COMBUSTION, Combustion Institute, Pittsburg, 675.

Harsha, P.T. (1973) AEDC-TR-73-177.

Heisenberg, W. (1949) *A. Physik 124*, 628.

Hanjalic, K. and Launder, B.E. (1972) *J. Fluid Mech. 52*, 609.

Hill, J.C. (1969) Goddard Space Flight Center, Preprint No. X-641-69-108, Greenbelt, Md.

Hilst, G.R. (1973) AIAA Preprint Paper 73-101.

Hopf, E. (1952) *Journal of Rational Mechanical Analysis 1,* 87.

Howe, N.M., Shipman, C.W. and Vranos, A. (1963) NINTH SYMPOSIUM ON COMBUSTION, Academic Press, 36.

Kolmogorov, A.N. (1941) *C. R. Academy of Sciences, USSR 30,* 301.

Kovasznay, L.S.G. (1956) *Jet Propulsion 26,* 485.

Kovasznay, L.S.G. (1962) THE MECHANICS OF TURBULENCE, Ed. A. Favre, Gordon and Breech.

Kovasznay, L.S.G., Kibens, V. and Blackwelder, R.V. (1970) *J. Fluid Mech. 41,* 283.

Kraichnan, R.H. (1959) *J. Fluid Mech. 5,* 497.

Kraichnan, R.H. (1962) PROC. 13th SYMP. APPL. MATH. 199, Amer. Math. Soc., Providence, R.I.

LaRue and Libby, P.A. (1974) *Phys. Fluids 17,* 1956.

Laufer, J. (1969) *AIAA J. 7,* 706.

Launder, B.E. and Spalding, D.B. (1972) MATHEMATICAL MODELS OF TURBULENCE, Academic Press, London.

Libby, P.A. (1973) NASA SP-321, 427.

Libby, P.A. (1973) *Comb. Sci. Tech. 6,* 23.

Libby, P.A. (1974) Paper presented at the AGARD Conference, Apr. 1-2, Liege.

Libby, P.A. (1975) *Comb. Sci. Tech.,* to appear.

Liepmann, H. and Laufer, J. (1949) NACA TN-1257.

Lin, C.-H. and O'Brien, E.E. (1972) *Astronautica Acta 17,* 771.

Lin, C.-H. and O'Brien, E.E. (1974) *J. Fluid Mech. 64,* 195.

Linan, A. (1963) Ph.D. Thesis, California Institute of Technology.

Lundgren, T.S. (1969) *Phys. Fluids 12,* 485.

Morkovin, M.V. (1962) *The Mechanics of Turbulence,* Ed. A. Favre, Gordon and Breach.

Mellor, G.L. and Herring, H.J. (1973) *AIAA Journal 11,* 590.

Murthy, S.N.B. (1974) (Editor) PROCEEDINGS OF THE SQUID WORKSHOP ON TURBULENT MIXING OF NON-REACTIVE AND REACTIVE FLOWS, MAY, 1974, to be published by Plenum Press, N.Y.

O'Brien, E.E. (1966) *Phys. Fluids 9,* 1561.

O'Brien, E.E. (1968) *Phys. Fluids 11,* 1883.

O'Brien, E.E. (1969) *Phys. Fluids 12,* 1999.

O'Brien, E.E. (1971) *Phys. Fluids 14,* 1326.

O'Brien, E.E. (1973) Paper presented at the SECOND IUTAM-IUGG SYMPOSIUM ON TURBULENT DIFFUSION IN ENVIRONMENTAL POLLUTION, Charlottesville, Va. See ADVANCES IN GEOPHYSICS Vol. 18B, F. N. Frenkiel and R. E. Munn (Editors), Academic Press, New York, 1974, 341.

Obukhov, A. M. (1941) *Compt. Rend. Acad. Sci. URSS 32,* 19.

Obukhov, A.M. (1949) *Compt. Rend. Acad. Sci. URSS 66,* 17.

Orszag, S.Q. (1971) THE STATISTICAL THEORY OF TURBULENCE, Cambridge Univ. Press, London.

Pao, Y.H. (1965) *Phys. Fluids 8,* 1063.

Patel, R.P. (1973) *AIAA J. 11,* 67.

Pearson, J.R.A. (1963) *App. Sci. Res. All,* 321.

Phillips, H. (1965) TENTH SYMP. INTERN. COMBUSTION, The Combustion Institute, Pittsburgh.

Rhodes, R.P. and Harsha, P.T. (1972) AIAA Paper No. 72-68.

Rhodes, R.P., Harsha, P.T. and Peters, C.E. (1974) *Acta Astronautica 1,* 443.

Rodi, W. and Spalding, D.B. (1970) *Warme-und-Stoffubertragung 3,* 85.

Rotta, J. (1951) *Z. Physik 129,* 547.

Rotta, J.C. (1971) AGARD CONFERENCE PROCEEDINGS, No. 93, London.

Scurlock, A.C. and Grover, J.H. (1953) FOURTH SYMPOSIUM ON COMBUSTION, Williams and Wilkins, 645.

Shackelford, W.L., Witte, A.B., Broadwell, J.E., Trost, J.E. and Jacobs, T.A. (1973) AIAA Preprint Paper 73-640.

Spalding, D.B. (1970) VDI-BERICHTE NR. 146, VDI VERLAG, Dusseldorf, 25.

Spalding, D.B. (1971) 13TH SYMPOSIUM (INTERNATIONAL) ON COMBUSTION, The Combustion Institute, Pittsburgh, 649.

Spencer, B.W. and Jones, B.G. (1971) AIAA Preprint Paper No. 71-613.

Stanford, R.A. and Libby, P.A. (1974) *Phys. Fluids 17,* 1353.

Tennekes, H. and Lumley, J.L. (1972) A FIRST COURSE IN TURBULENCE, The MIT Press, Cambridge, Mass.

Toor, H.L. (1962) *A.I.Ch.E.J. 8,* 70.

Townsend, A. (1951) *Proc. Roy. Soc. London 209A,* 418.

Townsend, A.A. (1956) THE STRUCTURE OF TURBULENT SHEAR FLOW, Cambridge University Press.

Way, J. and Libby, P. (1971) *AIAA J. 99,* 1567.

Westenberg, A.A. (1954) *J. Chem. Phys. 22,* 814.

Westenberg, A.A. and Rice, J.L. (1959) *Combustion and Flame 3,* 459.

Williams, F.A. (1971) ANNUAL REVIEW OF FLUID MECHANICS, Vol. 3, Annual Reviews, Palo Alto, Calif., 171.

Williams, F.A. (1974) Paper presented at the AGARD Conference, Apr. 1-2, 1974, Liege.

Winant, C.D. and Browand, F.K. (1974) *J. Fluid Mech. 63,* 237.

Wolfshtein, M.W. (1967) Ph.D. Thesis, Imperial College, London.

Wygnanski, I. and Fiedler, H.E. (1970) *J. Fluid Mech. 41,* 327.

Simulating Turbulent Field Mixers and Reactors

or

Taking the Art Out of the Design

G. K. Patterson

Chapter V

Simulating Turbulent-Field Mixers and Reactors
-or-
Taking The Art Out of the Design

G. K. PATTERSON

Department of Chemical Engineering
University of Missouri-Rolla
Rolla, Missouri

I. INTRODUCTION

Engineers would like to be able to design mixers and chemical reactors from a knowledge of basic hydrodynamic parameters, geometry and molecular properties. Molecular properties, such as viscosity, density, diffusivity and reaction rate constants, are not affected by the mixing process. The hydrodynamic parameters and geometry determine the intensity of the turbulence, the intensity of the turbulence determines the rate of mixing, and the rate of mixing, in many cases, determines the conversion within a reactor. If the reaction is very slow (low rate constant), the molecular reaction kinetics will be controlling and the rate of turbulent mixing will be unimportant. If the reaction is monomolecular or the reactants are premixed, the rate of mixing will affect the conversion only insofar as the reactants in one part of the reactor become diluted with product from another part of the reactor. This situation is called *backmixing*. For first-order reactions, the conversion in such a reactor is completely defined by the residence-time distribution (rtd). If the reaction is not first-order, the degree of backmixing and the rtd determine the conversion when one is dealing with premixed reactants.

If the reaction is bimolecular and the reactants are not premixed before entering the reactor, the rate of mixing between reactants will control the conversion if the reaction is molecularly fast (high rate constant). This situation could be called *reactant mixing*. To illustrate the difference between this situation and backmixing, one can consider the case of no mixing at the molecular level (i.e., no molecular diffusion). Backmixing will yield the highest possible conversion (for a particular rtd) for reactions of order greater than one; the reactant mixing case, however, will yield no reaction unless it can occur at interfaces between the segregated reactants.

The cases of no mixing at the molecular level and complete mixing to the molecular level are called *macromixing* and *micromixing*, respectively. Macromixing implies that fluid elements of a scale orders of magnitude larger than molecular scale may be mixed with one another, but no diffusion may take place between them. Micromixing implies that elements of fluid in proximity become completely mixed to molecular scale. Both macromixing and micromixing, as well as any intermediate level of mixing, may occur within any possible rtd. It is this important observation that makes it necessary to know more than the rtd for a reactor in order to determine the conversion for any but a first-order, monomolecular reaction.

The problem of determining distributions of intermediate levels of mixing and how the mixing interacts with reactions is the subject of this chapter. The degree of unmixedness is called the *intensity of segregation* and was first treated in relation to chemical reactors by Danckwerts (1958) and Zwietering (1959). They were concerned, however, with the backmixing discussed above and defined segregation in terms of the mixing of reactants and products. Here we are concerned primarily with the reactant mixing case, although backmixing cases will be mentioned for comparison. Segregation in the reactant mixing case is defined in terms of the mixing of two reactants (see Chapt. II).

II. TRADITIONAL DESIGN AND SCALE-UP OF MIXERS

Design and scale-up of turbulent mixers in the past have been done by using overall similarity concepts and rules of thumb such as power per unit volume, impeller torque and mixing time correlations in batch mixers as criteria for mixing extent. Progress has been made, however, in the application of numerical techniques to the calculation of transport processes in jet, wake and pipe flows (Hanjalic and Launder, 1972; Spalding, 1971). Such advances in practical knowledge and useful theoretical concepts of turbulence make it possible to consider mixing in a much more detailed manner. However, before a hydrodynamic model of mixing in vessels is presented, the older design and scale-up methods will be discussed.

Because of the presumed importance of scale-up methods for tanks based on impeller power per unit volume to some exponent, correlations of power consumption with vessel diameter and height; impeller diameter, type and shape; impeller rotation rate; fluid density and viscosity; and baffle arrangement have been extensively studied. A large body of this information for both propeller and turbine impellers has been reported by Rushton *et al.* (1950) and then summarized by Rushton and Oldshue (1953A, 1953B). Other major sources of data indicating power consumption as a function of the power number, $N_P = P/(\rho N^3 d_i^5)$, have been reported and discussed by Uhl and Gray (1966).

The relationship between power per unit volume and the slope of log Nusselt or Sherwood number versus log

Reynolds number was demonstrated for measurements with various size vessels by Rushton (1952). Later writers (i.e., Penny, 1971) have considered the various relationships between power per unit volume and vessel size for such desired results as constant bubble diameter, constant degree of solids suspension, constant heat transfer coefficients or constant blend times, for both laminar and turbulent mixing. In the case of high Reynolds number turbulent mixing scaled to achieve constant blend time, Penny (1971) recommended scaling power per unit volume with the square of the tank diameter. It will be shown later that this is compatable with a hydrodynamic turbulence model and forms the basis for a necessary assumption in forming a scale-up relationship.

In cases where uniform suspension of solid particles or of small bubbles is the important criterion for design, both constant impeller torque per unit volume and constant fluid velocities at equivalent locations for mixers of various sizes have been used for scale-up and design by Connolly and Winter (1969). It can be easily shown by dimensional arguments that constant torque per unit volume and constant flow velocity are equivalent scaling laws for turbulent mixing. The constant flow velocity basis for scale-up has also been recommended for dissolving, heat transfer and blending operations by Rushton (1952). There seems to be some justification in this recommendation for the surface to fluid transport processes such as heat transfer and dissolution, where convection coefficients are primarily velocity sensitive, but it will be shown later that the equal velocities criterion has little relation to blending (mixing) of miscible fluids.

Attempts have been made by Corrsin (1957, 1964) and Rosensweig (1964) to relate local mixing rate to local hydrodynamic parameters in order to gain more direct insight into the mixing process. Corrsin found relationships, starting with simple approximations to the turbulence energy spectra and the scalar spectra, which indicated that the normalized mixing rate is most strongly affected by a group containing the turbulence energy dissipation rate and a scalar length scale, ε/L_s^2. His analysis is presented in Chapt. II. If proportionality is assumed between quantities which are dimensionally equivalent for an entire mixer, then

$$P/\rho d_T^3 \propto \varepsilon$$

$$d_T \propto L_s$$

and

$$\tau^{-1} \propto (1/a'^2)(da'^2/dt)$$

Therefore

$$\tau^{-1} \propto (P/\rho d_T^3 d_T^2)^{1/3}$$

or

$$(N\tau)^{-1} \propto (P/\rho N^3 d_T^5)^{1/3}$$

If, then, mixers of equivalent blend intensity, $(N\tau)$, are desired, the power number, $(P/\rho N^3 d_T^5)$, is constant and $(P/d_T^3) \propto d_T^2$, or power per unit volume is proportional to volume to a 2/3-exponent. This matches the Penny proposal mentioned above.

Since for the effective blending of two miscible fluids, blend time or blend intensity is the criterion of performance, a number of investigators have performed experiments to determine the effects of variable parameters on that quantity. Fox and Gex (1956) and Norwood and Metzner (1960) each established similar relationships between various dimensionless groups and the Reynolds number:

$$(N\tau)(d_i/d_T)^n \Big/ (N^2 d_i/g)^{1/6} (H/d_T)^{1/2} = f(Nd_i^2\rho/\mu)$$

where n = 2 for Norwood and Metzner and n = 3/2 for Fox and Gex. The experiments used were batch dye dispersion tests. The first term may be set proportional to a term involving volumetric feed rate:

$$(N\tau) \propto (Nd_T^3/Q)$$

This relationship allows the results of dye disappearance tests to be related to the results of a model to predict extent of mixing in the effluent of mixers of various sizes and impeller speeds.

A concise summary of mixer scale-up relationships was given by Keey (1967). He discussed the geometric ratio factors and the exponents on these that have been advanced by Rushton *et al*. (1950) and others for scaling the effects of geometric changes on power number. Since there is little agreement on the values of many of the exponents, Keey concludes that such scaling is a risky procedure. Indeed, it seems that specific mixing information and reliable, tested calculation methods for several standard mixer configurations would provide a basis for a more reliable scale-up procedure. A hydrodynamic mixing model with established techniques for its application would help meet the second part of this prescription. Such a model would also present an opportunity of extension to mixers, such as jets, confined jets and pipe mixers, which are difficult to scale on the basis of power per unit volume and other related concepts.

III. DESIGN AND SCALE-UP OF TURBULENTLY MIXED REACTORS

The discussion so far has dealt with the design of mixers without consideration of parallel processes. Of even greater importance, however, is combined mixing and reaction in turbulent field reactors, primarily those without reactant premixing. Studies that involve the effect of the degree of mixing on reaction rate and, therefore, reactor size for given conversion can be placed into two general categories: 1) the *direct approach* whereby the reaction conversion is defined in terms of both the kinetic parameters (concentration, reaction rate constant, temperature and pressure) and the mixing parameters (concentration fluctuations, scale of turbulence and energy dissipation rates) and 2) the *indirect approach* whereby a model, often unrelated to the real physical mechanism, is constructed to estimate how far the reaction conversion will deviate from an ideal situation such as a perfect mixer. The direct approach could be called hydrodynamic mixing (hdm) models. It should be noted that, as is the case for most complex models of processes, an explicit expression for the size vessel required is not usually possible. Therefore, scale-up implies variation of size and other parameters until the desired result is obtained.

By far the bulk of the literature concerning chemical reactors utilizes the indirect approach because of the difficulty in obtaining turbulence data, and even more, because of the complexity of the mass, momentum and energy balance relationships for these systems. The indirect approach includes residence time distribution techniques combined with the concepts of micromixing and macromixing, dispersion and diffusion models, coalescence models and the two-environment models. Most of that work either ignores the effect of feed conditions or assumes premixed reactants being fed to the reactor. The rtd based techniques are amply reviewed by Levenspiel (1972) and Smith (1970).

Danckwerts (1953) initiated the use of distribution functions of residence time to calculate efficiencies of reactors and blenders. Later, Danckwerts (1958) presented the concepts of the degree of mixing and segregation for both the cases of non-premixed and premixed reactants when the reaction is second-order and of the form A + nB → Products. By premixed reactants, it is meant that species A and B are mixed thoroughly together on the molecular level before entering the reactor. If in addition to having premixed reactants, the fluid within the reactor is in the state of maximum mixedness (micromixed) as defined by Zwietering (1959), the entering feed is immediately mixed on a molecular scale with the entire age spectrum of molecules in the tank. Of if, in addition to having premixed reactants, the fluid within the reactor is macromixed, the entering feed is broken up into blobs or steaks which remain intact throughout their stay in the reactor. The micro- and macromixed cases are illustrated in Figs. 1 and 2.

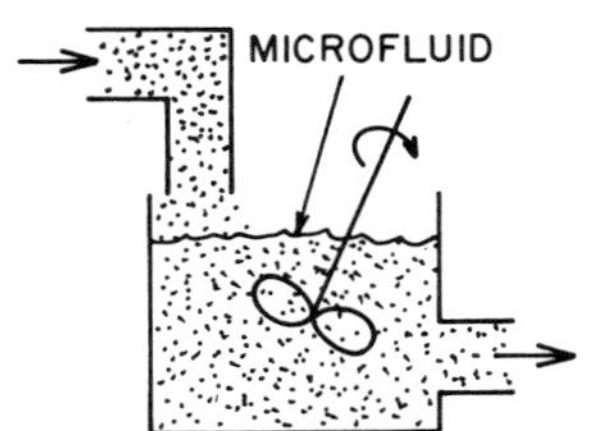

MIXING IS TO MOLECULAR SCALE-
SOME MIXING AT LARGER SCALES MUST OCCUR

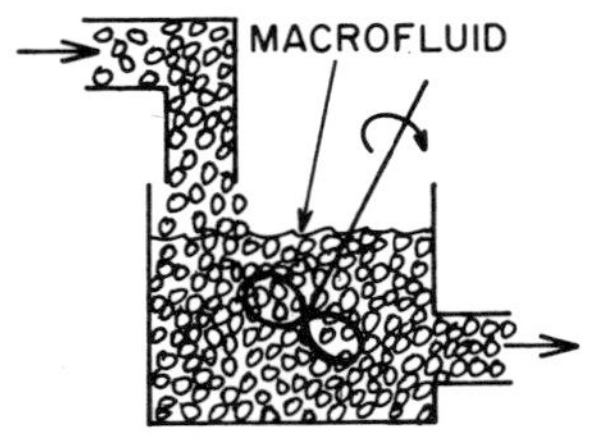

MIXING ONLY AT SCALES LARGER THAN MOLECULAR

Fig. 1. Schematics of micromixed and macromixed reactors with premixed feed.

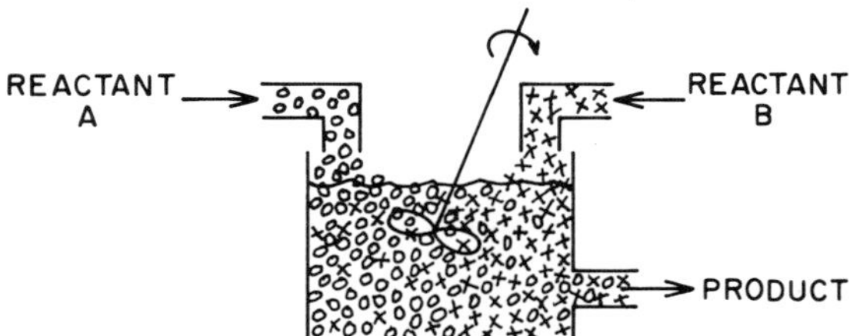

Fig. 2. Schematic of reactor with non-premixed feeds.

The micromixed case and the macromixed case are the extremes of mixing when the feed is premixed. Any real situation with premixed feed must lie between these extremes. Felder and Hill (1969) and Chauhan *et al.* (1972) showed for the case of premixed feed, that complete segregation (macromixed fluid) maximizes conversion for reactions whose order is greater than one and no segregation (micromixed fluid) maximizes conversion for reactions whose order is less than one. Methot and Roy (1971, 1973) studied the micro- and macromixed reactor models analytically in order to develop design equations and experimentally to determine how well they apply to premixed-feed reactors. The experimental results lay between the extreme cases. Levenspiel (1972) has provided design charts for both first- and second-order reactions for the micro- and macromixed cases, but only for stoichiometric feeds. Methot and Roy obtained conversion equations for arbitrary stoichiometric ratios, β. Plots of their equations in the form used by Levenspiel are shown in Figs. 3 and 4. In these

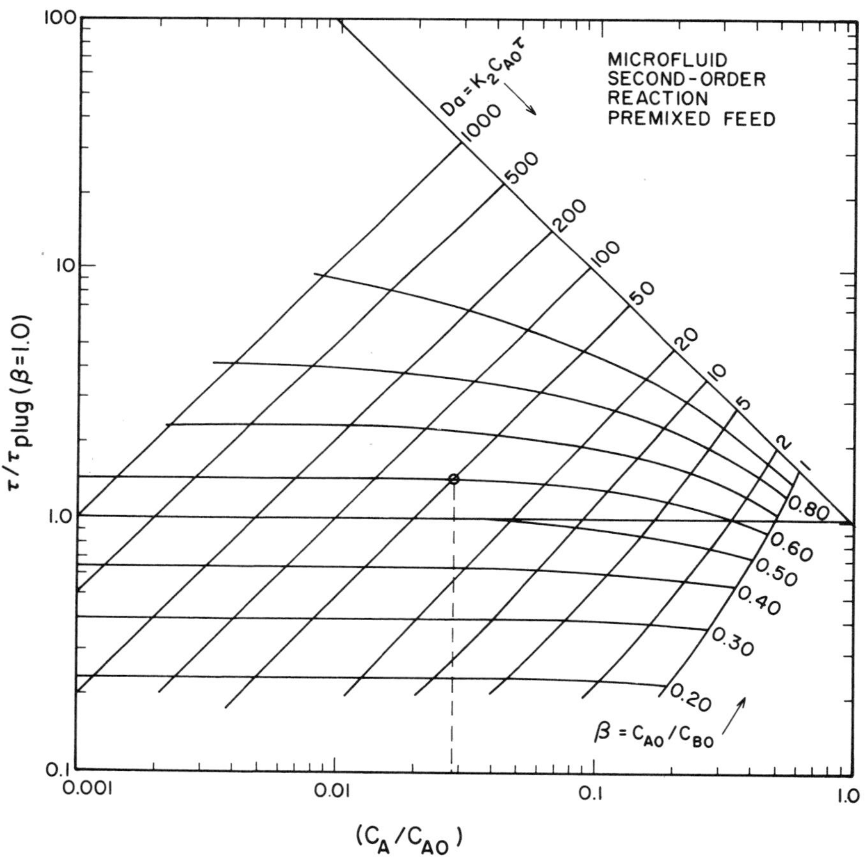

Fig. 3. Design chart for second-order reaction, premixed feed and various reactant ratios for micromixed reactants ($C_A = A$, $C_{AO} = A_O$, $C_{BO} = B_O$, $Da = N_D = kA_O\tau$).

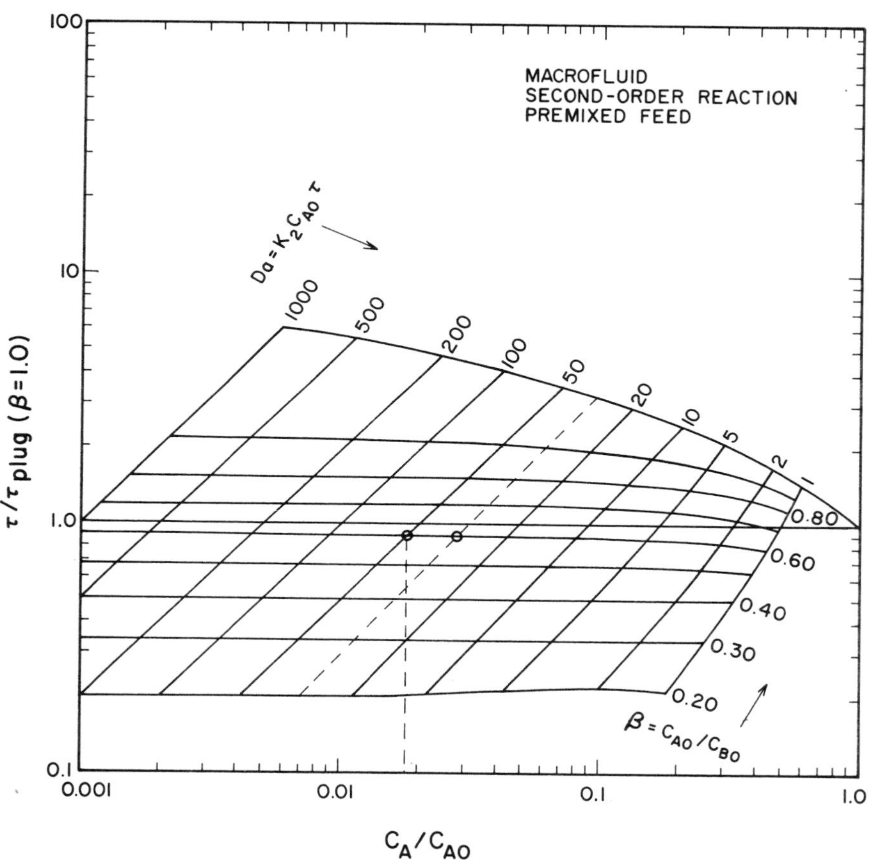

Fig. 4. Design chart for second-order reaction, premixed feed and various reactant ratios for macromixed reactants ($C_A = A$, $C_{AO} = A_O$, $C_{BO} = B_O$, $Da = N_D = kA_O\tau$).

charts, the necessary residence time for a given effluent conversion divided by the residence time of an ideal plug flow reactor for the same conversion is plotted versus normalized exit concentration with β and Damköhler number as parameters.

Takamatsu *et al.* (1971), Weinstein and Adler (1967), Keairns (1969), Treleaven and Tobgy (1971, 1972, 1973) and Kattan and Adler (1967) are other works which make use of the application of the extremes of mixing for the premixed-feed case to determine the effects of mixing on reaction. Leitman (1970) reviewed in great detail all the accepted rtd based methods of modeling and scaling-up reactors with premixed feeds.

When reactants A and B are not premixed before being fed to the continuous stirred tank reactor (cstr), the concepts of segregation have entirely different meanings than those for the premixed feed case. Complete segregation implies that blobs or streaks of unreacted A and B make up the contents of the reactor and no reaction can take place. On the other hand, no segregation implies that the diffusion process has reached equilibrium and the reactor contents are thoroughly mixed on a molecular scale and reaction can proceed at a maximum rate.

For the case of non-premixed reactants, Mao and Toor (1970) obtained reasonably good predictions of conversions measured by Vassilatos and Toor (1965) using a pure diffusion model for the effect of mixing on reaction (see also Chapt. III). Rao and Dunn (1970) used a random coalescence-dispersion (cd) model to also predict, with fair success, the conversion data of Vassilatos and Toor. The cd models simulate mixing by allowing the liquid at various sites (usually two at a time) to coalesce, react for a given time interval, then again disperse into the same number of sites. Random cd models are those in which the specified number of coalescing sites are chosen randomly. The Rao and Dunn model used a constant average rate of coalescences/site in the entire reactor. Canon *et al.* (1974) have applied the same cd model to the Vassilatos and Toor data but with variable rather than constant rate of coalescences/site in order to better account for variations in velocity and turbulence with length in the reactor. Results of these calculations for one stoichiometric ratio (β = 1.26) are shown in Fig. 5. Also shown are the results of an hdm model which will

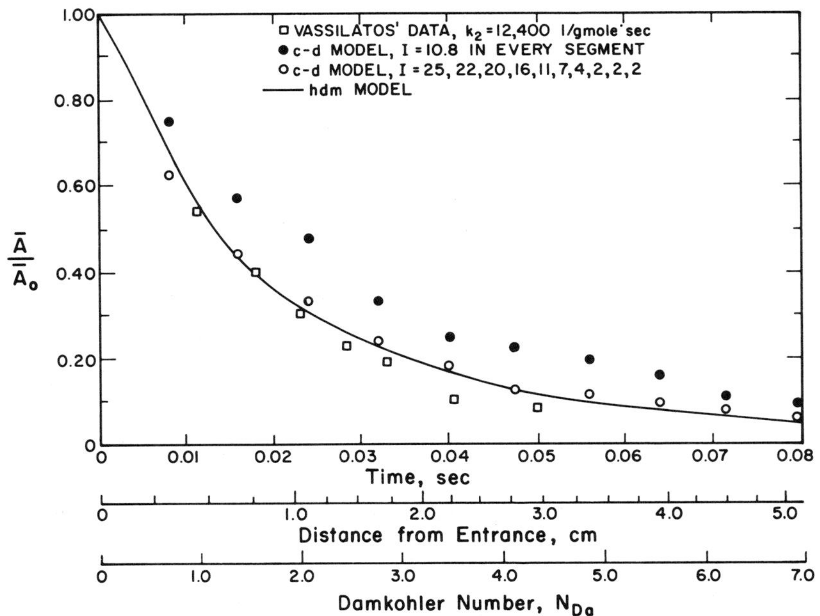

Fig. 5. Comparison of reactor mixing models with Vassilatos' data in a tubular reactor.

be discussed later. Rao and Edwards (1971) used their same cd model to calculate conversion for second-order, very rapid reactions in terms of concentration fluctuations and $\beta = B/nA$. The cd models of Spielman and Levenspiel (1965) and Kattan and Adler (1967) were applied to second-order reactions using both feed conditions. Levenspiel (1972) shows a design chart for stirred-tank reactors which gives the ratio of residence times at various coalescence rates as a function of conversion for zero- and second-order reactions. Damköhler number lines are superimposed. Zeitlin and Taylarides (1972) also calculated conversion-time profiles using a cd model for a second-order reaction taking place in the dispersed phase of a two-phase liquid-liquid agitated vessel with a premixed reactant feed.

Finally, the two-environment (te) models presented by Ng and Rippin (1965) to simulate intermediate levels of mixing in continuous flow systems is still another indirect approach to the prediction of reactor conversion. Other works specifically utilizing this type of model are by Nishimura and Matsubara (1970) and Rippin (1967). Actually, all the indirect models which attempt to account for intermediate levels of mixing may be considered to be te models since micromixing always takes place for some fraction of time or in some small region.

The Ng and Rippin te model makes use of one micromixed and one macromixed region with all feed entering the latter. The rate of material transfer from either environment to the effluent is proportional to the amount of material remaining in that environment. Rao and Edwards (1973) applied the Ng and Rippin model to cstrś for cases of premixed and non-premixed reactant feed and compared the results to results of calculations using models of Spielman and Levenspiel (1965)

Kattan and Adler (1967), and Villermaux and Zoulalion (1969). In the premixed cases, almost all results were identical for equivalent values of the mixing parameter (each contained one). In the non-premixed cases, each model yielded different results in detail, but trends with mixing parameter were similar. No comparisons with experimental results were made.

Another use of the indirect approach was that of Evangelista (1970) who developed a model for the backmixing effect in a reactor using the coalescence model of Curl (1963). An important parameter for correlating the effects of mixing in a premixed reactor, which resulted from Evangelista's work, is the mixing intensity, defined as ratio of the time constant for segregation decay to the residence time in the reactor. If the strongest parameter in the Corrsin equation for segregation decay is used, the mixing intensity becomes $I_E = (\varepsilon/L_s^2)^{1/3}/(V_T/Q)$. This parameter was used to correlate conversion predictions for first- and second-order premixed, parallel, consecutive and polymerization reactions. He also extended the Curl model to the case of non-premixed second-order reactions. Some results from Evangelista's and Spielman and Levenspiel's calculations for second-order premixed and non-premixed reactions are shown in Figs. 6 and 7. It should be emphasized that the conversion levels plotted against Damköhler number are mean values for the entire mixer and not effluent values obtained when the distribution of conversion is not uniform.

In contrast to the indirect modeling attempts described above, the direct approach used in this work is an attempt to relate the reaction rate directly to the turbulence parameters of the system which are in turn related to the type of reactor, flow patterns, agitation, etc. These relationships are mathematically modeled and solved simultaneously

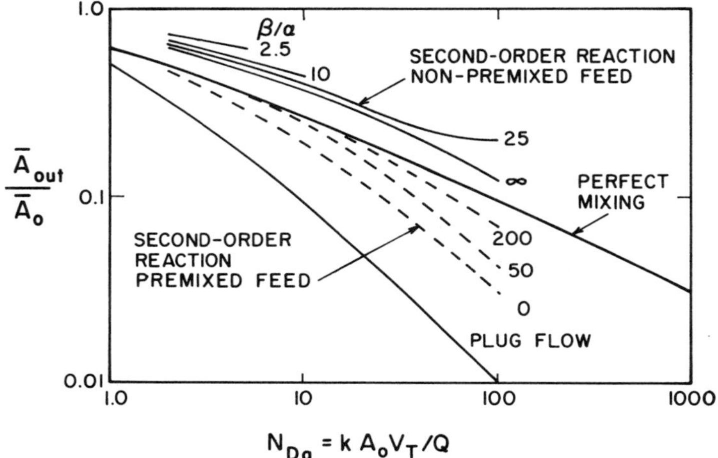

Fig. 6. Evangelista's model for mixing effects in a stirred-tank reactor, second-order reaction, premixed and non-premixed feed ($\beta/\alpha = I_E$).

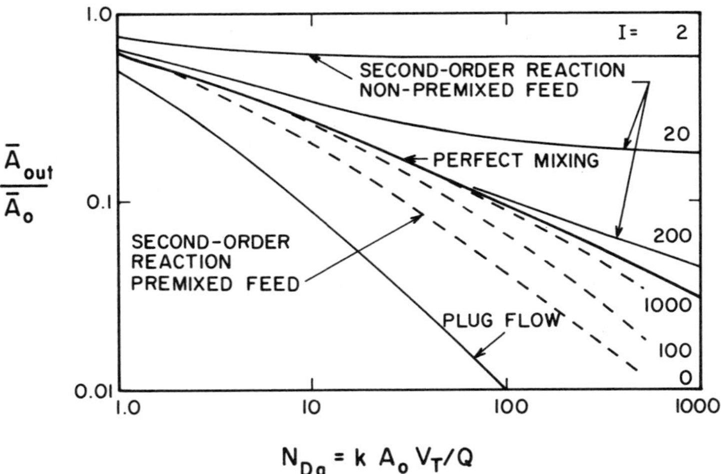

Fig. 7. Spielman and Levenspiel's model for mixing effects in a stirred-tank reactor, second-order, premixed and non-premixed feed.

to obtain reactor performance. All of the working parameters are at least in some manner related to physical properties of the system.

The literature that contributes to the direct approach includes that by Vassilatos and Toor (1965) who studied, in a tubular reactor, rapid and very rapid second-order reactions with no premixing of feed components. That work substantiated the theory developed by Toor (1962)concerning the effect of accomplished mixing on mean conversion and is discussed in detail in Chapt. III. The experimental data were used by Patterson (1970) for comparison with his hdm model. Toor (1969) and McKelvey *et al.* (1974) have also developed expressions and calculation methods for the second-order reaction rate in terms of time-averaged concentrations and segregation intensity and have demonstrated its validity for very slow and very fast reactions where reactant feeds are not premixed and when $\beta < 3$. This is also covered in detail in Chapt. III.

IV. DEVELOPMENT OF A HYDRODYNAMIC MIXING MODEL

The mathematical development of a hydrodynamic mixing (hdm) model based on modeling the effects of turbulence on mixing rate will be only summarized since the approach has been reported and discussed previously. Recent theoretical work on turbulent mixing includes that of Rosensweig (1964, 1966) who showed that the root-mean-square (rms) concentration fluctuation, integrated over a steady state mixing volume, may be treated as a conserved quantity. He used this concept in conjunction with his own expression for segregation decay rate to calculate the segregation intensity in the output of a uniformly mixed, stirred, baffled tank. *Uniformly mixed* means that the degree of segregation was uniform everywhere, obviously only a crude approximation to reality. Patterson (1970) used Rosenweig's conservation statement and the expression for the segregation decay rate developed by Corrsin (1964) in order to develop a hydrodynamic model of the mixing process in a stirred tank. The hdm model incorporated fundamental mixing parameters, such as turbulent energy dissipation rates, integral scales of turbulence, flow patterns, and internal velocities taken from the experimental work of Cutter (1966). Lee and Brodkey (1964A), Brodkey (1966), and Gegner and Brodkey (1966) demonstrated experimentally that Corrsin's decay rate expression gives an estimate for the segregation decay rate when a dye is injected into a turbulent pipe flow stream. Both the work of Rosenweig and Corrsin and, consequently, that of Patterson, dealt with a stirred-tank mixer.

Measurements of turbulence parameters in stirred tanks are varied, including the use of different sized equipment, measuring techniques, and measuring devices. The relevant work includes that of Kim and Manning (1964) who measured radial components of the turbulence energy and intensity spectra within a stirred, baffled tank with a pressure transducer probe. Later, Bowers (1965) measured tangential and vertical velocities along with turbulence intensity by means of vane and hot-wire anemometers. His results indicated that the ratio of turbulence intensity to the blade tip speed was not independent of scale for geometrically similar systems, even though the ratio of tangential or vertical velocity to blade tip speed was. Rao (1969) and Rao and Brodkey (1972) made similar measurements in the direction of the mean velocity using a hot-film probe. The mean velocity direction was determined by a three-dimensional pitot tube. Cutter (1966) made photographic measurements of mean and fluctuating components of the velocity of water in a fully baffled, stirred tank. He used simplifications of the Navier-Stokes equation and the energy equation to calculate angular momentum and flow of energy throughout the cstr, but concentrated mainly on the impeller stream. His data are used in the present work.

Some measurements of steady-state segregation intensity and other statistical concentration parameters are also available and include the work of Manning and Wilhelm (1963), Reith (1964), Christiansen (1969) and Becker *et al.* (1966). Manning and Wilhelm used a conductivity probe as did Reith, while Christiansen and Becker used light scattering techniques. Lee and Brodkey (1964B) and Nye and Brodkey (1967) developed a light transmission probe and it has been used by Rao and Brodkey (1972) to measure transient concentration fluctuations in the impeller stream of his stirred tank. Steady-state measurements could not be made because the mixed fluid reduced light transmission to low levels.

V. APPLICATION OF THE HYDRODYNAMIC MIXING MODEL TO REACTORS

The work presented here is an extension of earlier efforts to model the effect of mixing on second-order reactions in a stirred-tank reactor with unmixed feed reactants. Such a model was first presented by Patterson (1970) and he related the degree of mixing to the reaction rate (or level of conversion) as a function of position in the reactor. These results were obtained through the simultaneous solution of a set of segregation balance equations. The balance equations were later improved by Otte *et al.* (1971) as indicated by comparisons of calculated results with the tubular reactor data of Vassilatos and Toor (1965).

For the case of second-order, irreversible reaction

kinetics, the following equation gives the rate of disappearance of B:

$$-\partial B/\partial t = kAB \qquad (1)$$

If concentrations of both reactants are independently variable in space due to the mixing process and $A = \overline{A} + a$ and the same for B, then

$$-\partial\overline{B}/\partial t = k(\overline{A}\ \overline{B} + \overline{ab}) \qquad (2)$$

after time-averaging, as has been pointed out in previous chapters. Toor (1969) has shown that the correlation $\overline{ab}$ may be represented by $-b'^2/\beta$, if the reaction rate is infinite and the reactants are not premixed. The quantity b' is a measure of the intensity of the segregation of B from A in the reactor. An equation for the rate of decrease of B due to turbulent mixing, molecular diffusion, and chemical reaction may be derived from the continuity equation for B which has been presented in previous chapters. The resulting equation is

$$(\partial b'^2/\partial t) + \overline{\vec{U}}\cdot\vec{\nabla}b'^2 = D\{\nabla^2 b'^2 - 2\overline{(\vec{\nabla}b)^2}\}$$

convection of a' — decay of a' by mixing

$$-2k(\overline{B}\ \overline{ab} + \overline{A}b'^2 + \overline{ab^2}) \qquad (3)$$

decay of b' by reaction

where $\vec{U}$ and $\vec{\nabla}b^2$ were assumed uncorrelated.

Again for infinite reaction rate, $\overline{ab}$ may be approximated by b'^2/β, and $\overline{ab^2}$ may be assumed to be zero. * From

---

*Of course, infinite molecular reaction rates cannot be used in this model since the last term becomes indeterminant ($k \to \infty$ as $\overline{B}\ \overline{ab} + \overline{A}b'^2 \to 0$, but rate constants up to 10^7 *ℓ/mole sec* have been used.

comparisons (Patterson, 1973) of steady-state model results with the data of Vassilatos and Toor (1965) for second-order reactions of varying β-values, the best approximation for $\overline{ab}$ at intermediate and slow reaction rates was found to be $-b'^2 (1 - \gamma)/\beta(1 + \gamma)$, where $\gamma = (\overline{B}^2 - \beta b'^2)/(\overline{B}^2 + \beta b'^2)$. Here γ represents the degree of interdiffusion of the reactants.

A continuity equation for b'^2 may be written as

$$\partial b'^2/\partial t + \vec{\overline{U}} \cdot \vec{\nabla} b'^2 = r_B \qquad (4)$$

When Eq. (4) is combined with Eq. (3), assuming $\nabla^2 b'^2$ is zero as per the former condition on averaged quantities, one obtains

$$r_B = -2D\overline{(\vec{\nabla} b)^2} - 2k\{Ab'^2 - \overline{B}b'^2(1 - \gamma)/\beta(1 + \gamma)\} \qquad (5)$$

Corrsin (1964) derived an expression for the rate of decay of segregation with high Schmidt number in isotropic turbulence which will be used as an approximation to the second term in Eq. (5). Substitution yields

$$r_B = -2b'^2/\{4.10(L_s^2/\varepsilon)^{1/3} + (\nu/\varepsilon)^{1/2}\ell n\, N_{Sc}\} - 2k\{Ab'^2 - \overline{B}b'^2(1 - \gamma)/\beta(1 + \gamma)\} \qquad (6)$$

In order to use Eqs. (2) and (6) to determine the distributions of $\overline{B}$ and b' in the reactor, balance equations for a given reactor subdivision (j) of volume V_j must first be written:

$$V_j(\partial\overline{B}/\partial t) = \sum_i Q_i\overline{B}_i - \sum_o Q_o\overline{B}_j + V_j R_{Bj} \qquad (7)$$

$$V_j(\partial b'^2/\partial t) = \sum_i Q_i b_i'^2 - \sum_o Q_o b_j'^2 + V_j r_{Bj} \qquad (8)$$

where i refers to input streams and o to output streams. The rate terms $V_j(\partial\overline{B}/\partial t)$ and $V_j(\partial b'^2/\partial t)$ would be zero for steady-state conditions.

Equations (7) and (8), when combined with Eqs. (2) and (6), describe the time-averaged concentration and the time-averaged concentration fluctuations (segregation) in a sub-volume (segment) of reactor. These two equations provide a mathematical model for the mixing process with second-order reaction occurring. A study of the model requires the solution of these two coupled, non-linear, first-order differential equations in $\overline{b'^2}$ and $\overline{B}$.

Application of the hdm model to reactions of order different from two is only beginning. Interest was at first strongest in that application because of the unsolved nature of the second-order non-premixed case. First-order reactions are, of course, not of interest here because the level of mixing does not affect their rates, only the residence time distribution. Fractional orders are difficult because of the complexity that results in the term for decay of b' by reaction in Eq. (3) and in the kinetic relationship of Eq. (2). Direct numerical approximations to these terms are being attempted for both premixed and non-premixed cases.

The premixed second-order case requires a different approximation to $\overline{ab}$ from both Eqs. (2) and (3). Toor (1969) has shown that for the infinitely slow case $\overline{ab} = + \overline{b'^2}/\beta$. This approximation should apply to premixed reactions up to moderate rates. Since fast reaction *cannot* be premixed, such cases are of no importance.

Heat transfer and generation in the reactor may be accounted for by balancing the flow of energy into and out of an element of volume with the energy generated. This energy balance can be written as follows:

$$\sum_i Q_i(\rho C_p T)_i = \sum_o Q_o(\rho C_p T)_o - \Delta H_R(\sum_i Q_i \overline{B}_i - \sum_o Q_o \overline{B}_o) + h_w A_{Rj}(T_j - T_w) \tag{9}$$

where i and o have the same meanings as before, and w refers to the wall. The reaction rate constant in the material balance calculations made here was allowed to depend on T_j:

$$k = k_0 \exp - \frac{\Delta E}{R}\left(\frac{1}{T_j} - \frac{1}{T_0}\right) \tag{10}$$

where k_0 equals the rate constant at T_0 and T_j is the temperature in segment j.

VI. COMPUTING-FLOWSHEET FOR THE MODEL

Figure 8 shows in series and parallel blocks how the hdm model for mixing effects in reactors is used. It is not intended to imply that all the variations of geometry shown are already included in the model program. So far, only the second-order, irreversible reaction has been studied for both tank configurations, the tubular plug flow, and in a preliminary manner, the confined jet. Actually, the same computing blocks are used to establish the basic variables used to calculate the coefficients in the balance equation matrices. It is simply a matter of setting coefficients within the computing blocks to zero or some positive or negative value to suit the geometry involved. Once the segregation, material and energy balance matrices are established, convergence to a solution is a general routine.

VII. APPLICATION OF THE MODEL TO TANK-MIXER REACTORS

In order to apply the hdm model to tank mixers with impellers, it is necessary to segment the tank so as to obtain a numerical approximation of the distribution of segregation over the tank. The balance equation applied to each of these segments requires a knowledge of the flow pattern in the tank and the local mixing rate. In the most general case, the velocities, rate of turbulence energy dissipation and length

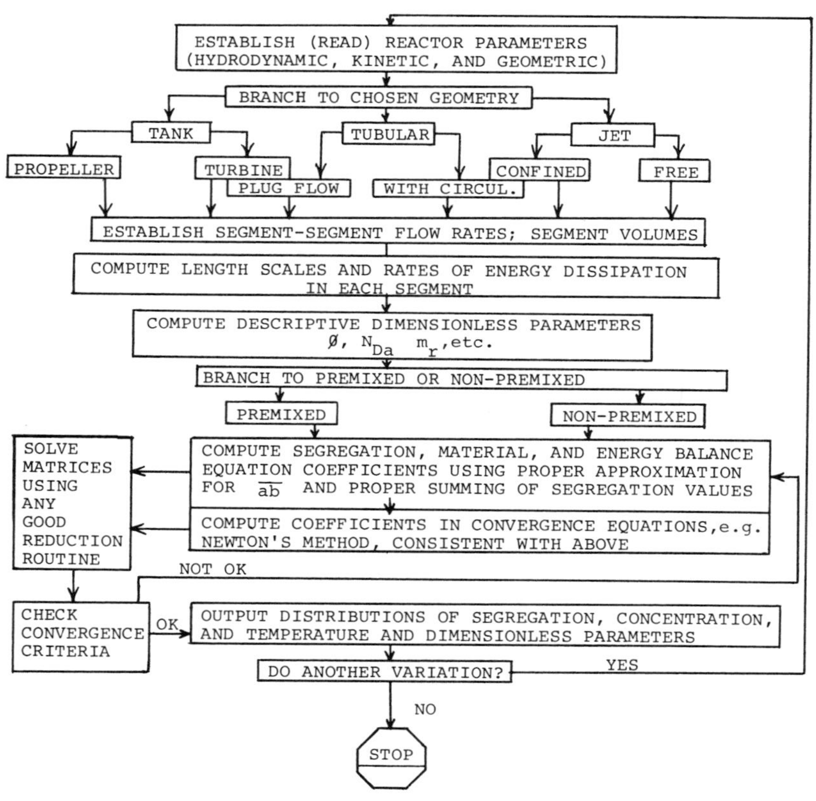

Fig. 8. Computing flow chart for HDM model.

scale for the local mixing rate could be obtained by use of a hydrodynamic model for turbulent flow, but since the circulating flow involved in tank mixers is extremely complex and difficult to model, flow rate (Q_{ij}), rates of turbulence energy dissipation (ε_i) and length scales (L_i) were estimated from experimental measurements.

For the turbine driven mixers, the experimental information was taken mainly from the measurements of Cutter (1966), who measured velocities, scales and rates of energy dissipation primarily in the impeller stream region. Cutter's tank was approximately 30 cm x 30 cm, enclosed, with a centered 10 cm diameter x 3 cm wide turbine impeller and four baffles which protruded 3 cm into the tank. Mean velocities and other hydrodynamic parameters have been measured by other investigators, e.g., Rao and Brodkey (1972), but the measurements by Cutter seem to be self-consistent and as complete as any. When Cutter's energy dissipation rates are integrated over the entire tank, the expected impeller power from power number correlations is obtained. By use of continuity considerations, overall flow patterns with velocity as a function of impeller rotation rate were deduced from the impeller stream information. The volumetric flow rates were determined for each segment for a thirty segment tank model as illustrated in Fig. 9. The rate of energy dissipation (ε) and scale (L_s) in each segment was also established. The modeled tanks were maintained geometrically similar at each size (30 cm, 90 cm and 210 cm in diameter). The correlation of Norwood and Metzner (1960) was assumed valid for small changes in geometric ratios.

The propeller driven mixers modeled had downflow impellers of standard three-blade design centered in the tank. The flow pattern information was derived from circulation

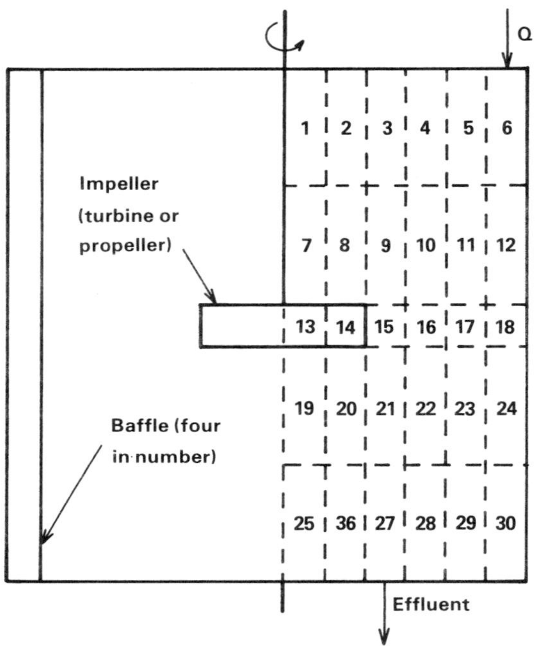

Fig. 9. Tank mixer for HDM model.

rate measurements made by Porcelli and Marr (1962). Volumetric flow rates were determined for each segment. Since Porcelli and Marr did not attempt to measure rates of energy dissipation and length scales, it was necessary to deduce these from other information. The total rate of turbulent energy dissipation must be nearly equal to the power used to drive the impeller, which for a standard propeller mixer may be obtained from the power number correlations of Rushton *et al.* (1950) and others. In this case, it was assumed that $P \simeq 0.32\rho N^3 d_i^5$ and that half the total dissipation is in the impeller stream below the impeller. This is consistent with findings for the turbine impeller (Cutter, 1966; Rao and Brodkey, 1972). Length scales were estimated from a relationship proposed by Rao and Brodkey (1972), $L_s = 3/4 (d_i/4)$, at the impeller and gradually larger away from the impeller. It must be emphasized that more experimental work in this mixer type is necessary in order to refine the estimates of these parameters. As in the case with the turbine mixer, the geometry of the propeller mixer was maintained similar at all sizes modeled. In all cases for both mixers, water was the model fluid with $\rho = 1.00$ g/cc, $\mu = 1.00$ cp, and $D = 0.00013$ cm^2/sec. Effects of changes in these properties have not been investigated.

The scale-up of each of the relevant parameters was based on the following assumptions for both tanks:

$$L_i \propto d_T$$

$$\varepsilon_i \propto d_T^2 \; *$$

$$Q_{ij} \propto N d_T^3$$

These assumptions represent application on a segmental basis of scaling laws which are approximate for entire mixers. Impeller rotation rates, N, modeled were 200, 360, 400 and 600 rpm. (When the impeller was a propeller, 1800 rpm was used for some reactor cases.) In all cases, the feed fluids were assumed to be completely unmixed and so the degree of segregation of the feed was calculated as follows:

$$b'^2 = \overline{A}_0 \overline{B}_0$$

where $\overline{A}_0$ and $\overline{B}_0$ were concentrations of species A and B after mixing if no reaction took place.

---

*consistent with dimensional considerations

VIII. RESULTS OF APPLICATION OF THE MODEL TO MIXING WITHOUT REACTION

Application of the model to the turbine and propeller mixers with the size and rotation rate variations mentioned previously and with feed locations at the top (segment 6) and at the impeller center (segment 13) resulted in distributions of segregation for each condition studied. As examples, the flow patterns and the segregation distributions for both types of mixers are given for one condition of each type in Figs. 10 and 11. In each case, relative volumetric

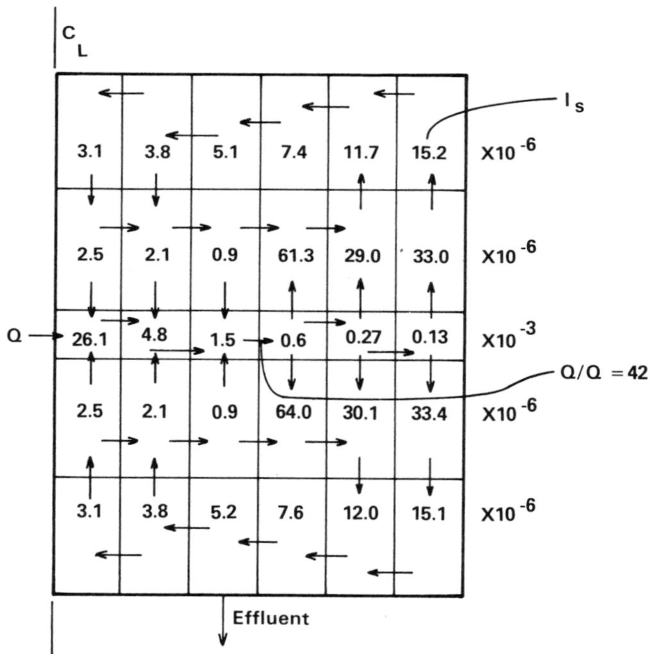

Fig. 10. Volumetric flow rates and segregation distribution in a turbine mixer with feed at the impeller center, $\phi = 0.015$ (Q/Q is local flow ratio to inlet flow).

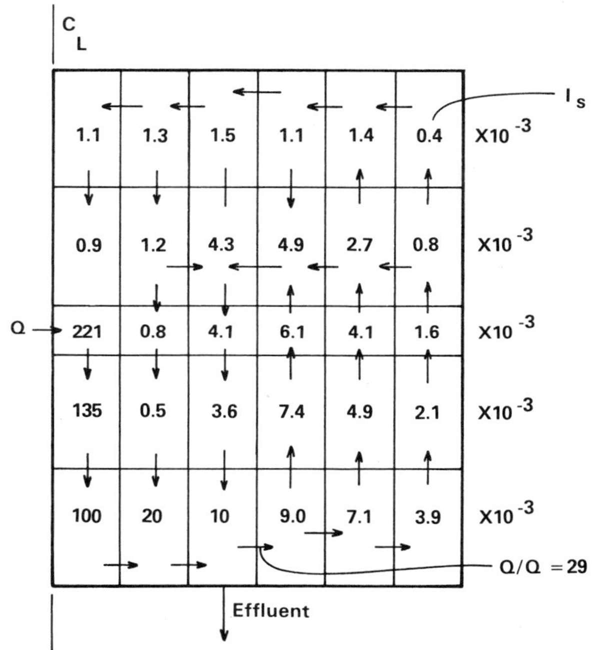

Fig. 11. Volumetric flow rates and segregation distribution in a propeller mixer with feed at the impeller center, ϕ = 0.015 (Q/Q is local flow ratio to inlet flow).

flow rates are indicated by arrow length and segregation as I_s. Plots could be made showing the variation of segregation at the effluent of the mixers, but a general representation based on the use of the dimensionless groups of Norwood and Metzner is much more compact and reveals as much information. Figure 12 shows all the effluent segregation values computed for both mixer types and both feed locations as a plot of the outlet I_s against

$$\phi = (Q/Nd_T^3)(d_T/d_i)^2(d_T/H)^{1/2}$$

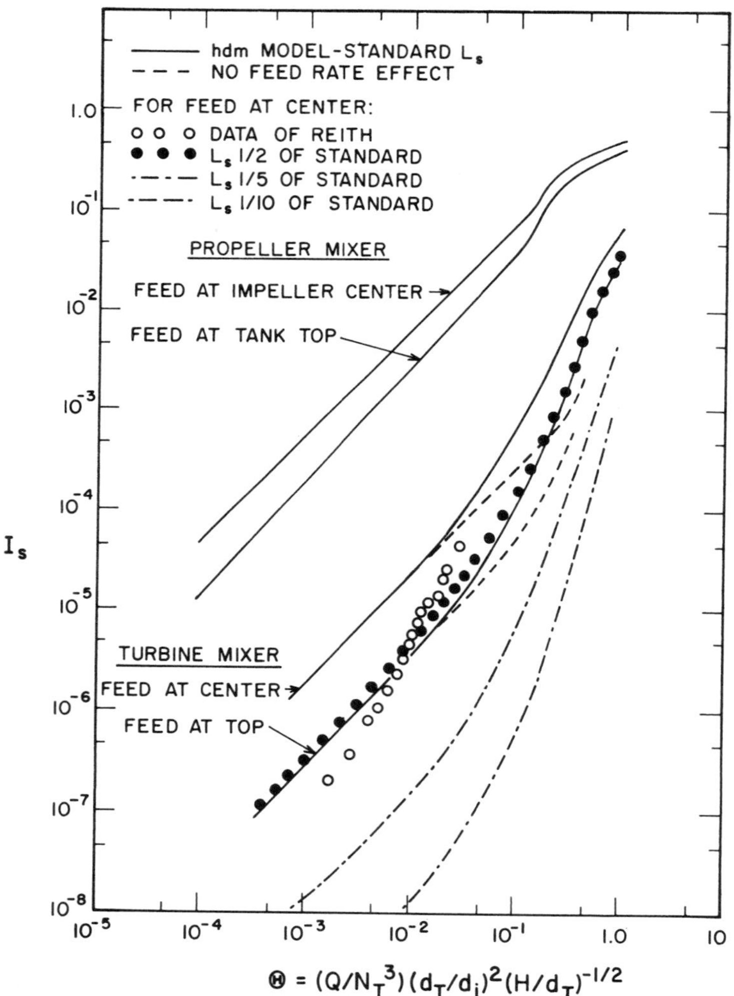

Fig. 12. Correlation of mixer output segregation for the HDM model compared to the data of Reith (x- axis is $\phi = (Q/Nd_T^3)\ldots$).

For each of the two feed points in each of the two mixer types, all the data fell upon a single curve with no effect of tank size or impeller rate when a given feed momentum effect on circulation rate and a given L_s distribution were used. The four solid line curves resulted for eventual use as scale-up charts for the four different geometries. The I_s variations caused by feed point variations are not as strong as the difference between the two mixer types.

The set of dashed lines illustrates the sensitivity of the model to feed momentum (jetting). There is a strong effect for both feed points in the turbine mixer in the intermediate feed rate range, but almost no effect in the propeller mixer. Somewhat greater effects might have been seen if the feed jetting had been associated with lower scales and high energy dissipation rates. The dotted lines represent the effect of changes in scalar length scale as indicated. The smaller length scales cause lower effluent segregation with greater slope of I_s with feed rate.

A comparison of the data of Reith (1964) (30 cm x 30 cm turbine mixer with d_i/d_T = 0.25 and feed at the impeller center) with these calculations is also shown in Fig. 12. Reith measured concentration fluctuations in mixing a salt solution with water by using conductivity probes. The correspondence between the model calculations and the experimental results is good for L_s values of approximately one-half the standard values. The smaller scale is probably necessary because the salt solution stream to Reith's mixer was at such a low rate that very small scalar lengths were probably developed.

The comparison between the calculation and the experiments of Reith shown in Fig. 13 for the rate of segregation decay in the impeller stream shows a somewhat lower

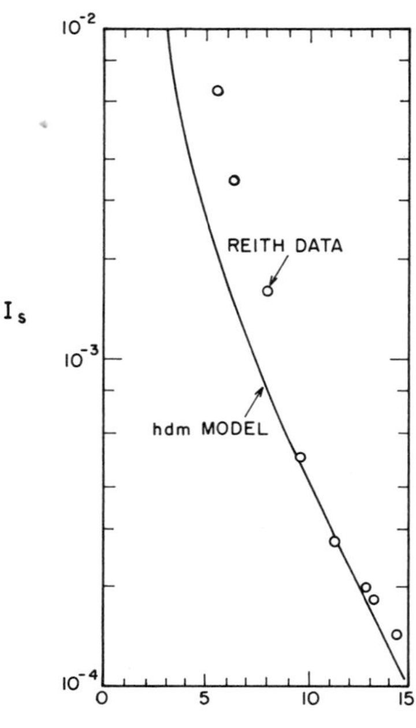

Fig. 13. Comparison of the HDM model results with the data of Reith in the impeller stream for 360 rpm, tank diameter of 30 cm, and ϕ = 0.015.

slope for the model, but none the less an excellent correspondence.

Unfortunately, no detailed experimental data exist for propeller mixed tanks, but comparisons of mixing times between propeller mixers and turbine mixers have been made. The dye disappearance data of Prochazka and Landau (1961) shows $(N\tau)_{propeller}/(N\tau)_{turbine} \simeq 1.87$ for $d_T/d_i = 4$ in both cases. Marr and Johnson's (1963) data indicate $(N\tau)_{propeller}/$

$(N\tau)_{turbine} \simeq 2.19$ for $d_T/d_i = 3$. An equivalent comparison in continuous flow tanks would be $(Q/Nd_T^3)_{propeller}$, since $d_T^3/Q \propto \tau$. Comparisons at equal values of Q/Nd_T^3 from the model calculations (values of effluent I_s in Fig. 12) give ratios of nearly 100. The relevant comparison, however, is actually the mean of the segregation values in the entire tank since that is what dye disappearance most likely measures. When conditions are found giving equal averages, the (Q/Nd_T^3) ratio for turbine to propeller mixers is about 2.44, indicating approximately correct behavior.

The dashed curves in Fig. 12 show the effluent levels of I_s obtained when no effect of feed rate is included in impeller stream flow rates. Since all detailed studies of mean velocities in mixers have been for no-feed operation or for very low feed rates as, for example, by Rao and Brodkey (1972) who used $(Q/Nd_T^3)(d_T/d_i^2)(d_T/H)^{1/2} \simeq 10^{-2}$, there is no experimental evidence of the detailed effects of high feed rates on flow pattern.

IX. RESULTS OF APPLICATION OF THE MODEL TO SECOND-ORDER REACTION

The model was applied to mixing and reaction, with reaction rate constants and feed rates varied such that a range of Damköhler numbers from 25 to 2.5×10^7 was covered. Reactor size was varied from 30 cm x 30 cm to 210 cm x 210 cm. The same range of impeller speeds was covered, except for some calculations for turbine mixed reactors at 50 rpm and for propeller mixed reactors at N = 1800 rpm. Much of the information summarized here, as well as temperature distribution results not discussed in this paper, has been reported comprehensively by Otte (1974).

The result of the calculation of reactor conversion

with the model is a distribution of conversion and segregation over the entire reactor. As an example, Fig. 14 shows distributions for one turbine mixed reactor condition and an

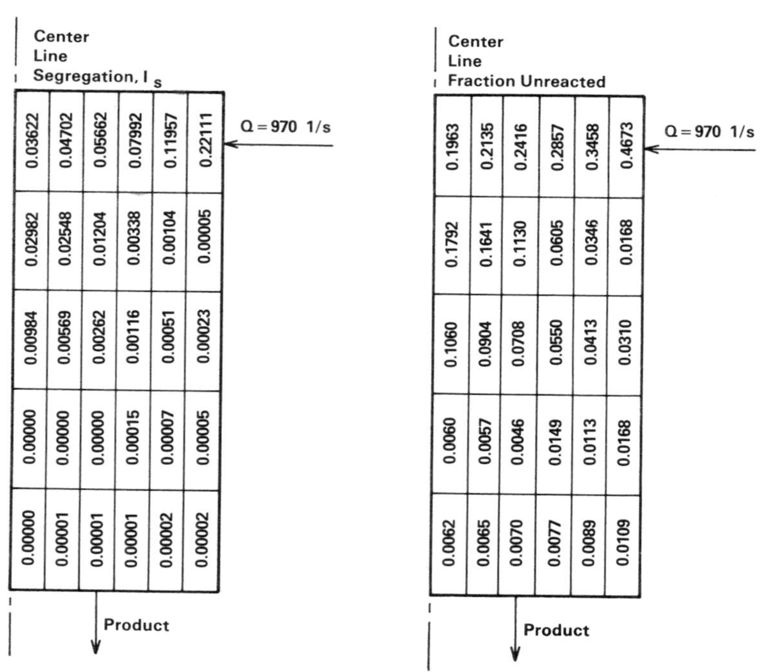

Fig. 14. Distribution of conversion in a turbine mixed reactor for 600 rpm, tank diameter of 210 cm, N_D = 2450, and m_r = 0.755.

isothermal second-order reaction (210 cm x 210 cm reactor size, 600 rpm impeller speed, N_D = 2450, $m_r = (\varepsilon/L_s^2)^{1/3}_{seg.}\ 13/(k\bar{B}_0)$ = 0.755 (see below), ϕ = 0.090, non-premixed reactants with a ratio of one), and Fig. 15 shows the same distributions for a propeller mixed reactor with all other conditions the same except that m_r = 0.063. As illustrated by these distributions, the propeller mixed reactor tends to be more uniform at the same impeller speeds. This is even more so at the same

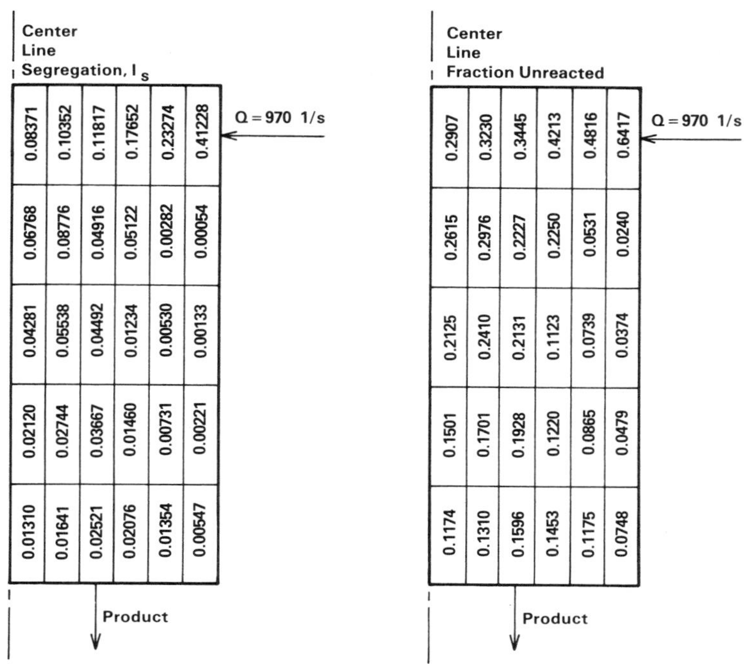

Fig. 15. Distribution of conversion in a propeller mixed reactor for 600 rpm, tank diameter of 210 cm, N_D = 2450, and m_r = 0.063.

input power levels, corresponding to equal m_r - values which require much higher propeller speeds.

The results of many different computations for isothermal non-premixed second-order reactions are shown in Figs. 16 and 17. Plotted are the normalized effluent concentrations (1 minus conversion), for feed at the top and product at the bottom for turbine and propeller mixed tanks, respectively, versus $N_D = k\bar{B}_0 V_T/Q$, the Damköhler number. The other parameter on the plots is the relative mixing intensity m_r, which is not the same as Evangelista's I_E. Here, we have the ratio of mixing rate to reaction rate, instead of mixing rate to space velocity.

For the turbine mixed reactor, each curve of constant relative mixing intensity begins (or would begin at low enough N_D) above the line for a perfectly mixed reactor. Actually, all these lines must approach $\overline{A}/\overline{A}_0 = 1.00$ for very low N_D. The curves of constant m_r approach then drop below the perfectly mixed line. They then, at some point, curve back up above the perfectly mixed line to seek an asymptotic deviation above it. The descending portion of each m_r curve represents a mixing controlled situation since the γ-values (see Eq. 6) are always small (<0.1). The ascending portions of the m_r curves represent reaction rate controlled behavior, since the γ-values are very nearly 1.0 (>0.9). There is, of course, a transition region at intermediate values of N_D. At higher N_D -values, the process becomes even more reaction rate controlled and the degree of deviation above the perfectly mixed case is merely a function of m_r. The rather complex crossings of the m_r curves is caused by complex interactions between the local turbulent mixing and the circulation effects on conversion.

For comparison sake, some lines of constant m_r for premixed feed are shown in Figs. 16 and 17. For these computations, each segment of the reactor was assumed perfectly mixed since, in premixed cases, segregation is generally low and the process is reaction rate controlled. The $\overline{A}/\overline{A}_0$ values in the effluent behaved similarly for the turbine and propeller mixed cases. Of course, as one would expect, the higher the circulation rate (higher m_r), the more like perfectly mixed was the reactor.

Table I lists comparisons of effluent concentrations for the turbine-mixed hdm model with those (which are overall

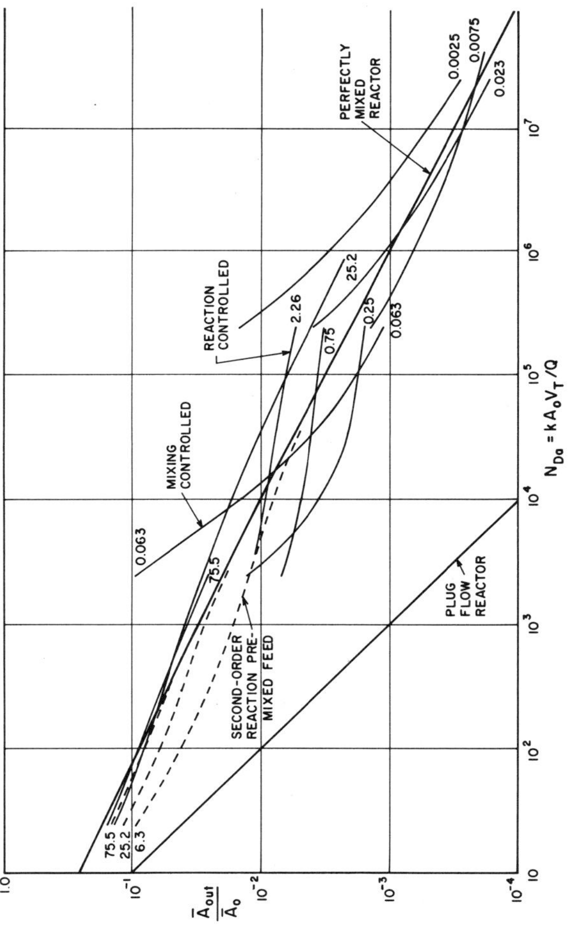

Fig. 16. Conversion as a function of Damköhler number and relative mixing intensity for turbine mixed reactors (parameter is m_r evaluated in seg. 13).

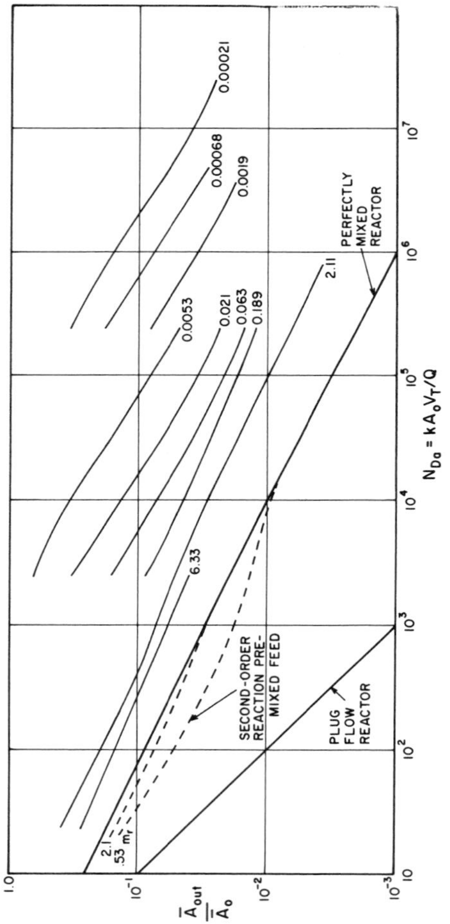

Fig. 17. Conversion as a function of Damköhler number and relative mixing intensity for propeller mixed reactors (parameter is m_r evaluated in seg. 13).

TABLE I *Comparison of Reactor Models* (N = 400 rpm and $B_0/A_0 = 1$)

REACTOR CONDITION	NORMALIZED EFFLUENT CONCENTRATION							
N_D	Perfectly Mixed Reactor	Turbine Non-premixed HDM Model $m_r = 50.4$	Ng and Rippin 2-Environment Model R = 100	Spielman & Levenspiel C-D Model I = 200	Evangelista Non-premixed Model $I_E = 50$	Evangelista Premixed Model $I_E = 50$	Turbine Premixed HDM Model $m_r = 50.4$	Methot & Roy Macro-Mixed Model
10,000	0.010	0.016	0.0199	0.038	--	--	0.010	0.0007
1,000	0.030	0.038	0.041	0.038	--	--	0.029	0.006
100	0.090	0.085	0.104	0.108	0.175	0.120	0.081	0.040
10	0.250	0.20	0.269	0.27	0.383	0.220	0.190	0.200

These models assume non-premixed feeds; degree of mixing and conversion is uniform in the reactor.

These models assume premixed feeds; degree of conversion is uniform in the reactor.

averages in the entire reactor) for the Ng and Rippin (1965), the Spielman and Levenspiel (1965), the Evangelista (1970) and the Methot and Roy (1971) models. The good correspondence between the Ng and Rippin te model at $R\tau = 100$ and the hdm model is noteworthy. R is the transfer rate between the macro- and micromixed regions. The te model predicts reactant concentrations slightly higher than the effluent concentrations of the hdm model since the hdm model accounts for the concentration gradients in the reactor. The Evangelista model does not agree well with the others. The results of the hdm premixed model deviate strongly from the macromixed model since perfect mixing was assumed in every segment.

The propeller mixed reactors did not show the dip of the constant relative mixing intensity curves below the perfectly mixed line. The γ-value in all cases was small; therefore, the reaction process remained mixing controlled. The complex interaction between local mixing effects and circulation effects does not seem to occur in propeller mixed reactors, presumably because the local turbulent mixing rate does not outstrip the circulation capabilities (tendency to make the reactor uniform) in a propeller mixed reactor as it does in a turbine mixed reactor.

X. COMPARISON OF MODEL WITH EXPERIMENTAL DATA FOR A TUBULAR REACTOR

There seem to be no published reactor conversion data for cstr's of standard design with unmixed feed. We attempted, therefore, to refine the hdm model through comparisons with the non-premixed feed tubular reactor results of Vassilatos and Toor (1965).

The tubular reactor of Vassilatos (1964) was modeled by a single line of small subdivisions (cells) as

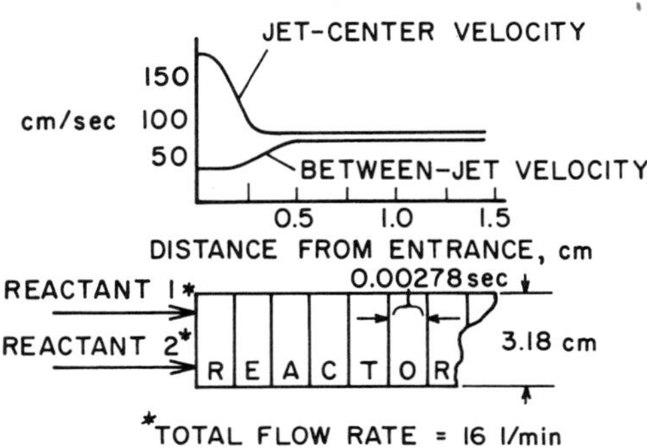

Fig. 18. Tubular reactor of Vassilatos.

shown in Fig. 18. The comparison calculations were made for a feed rate of 16 ℓ/m, the lower of two feed rates Vassilatos used in his reaction conversion measurements. The reactor used and modeled here had an inside diameter of 3.18 cm with injection of each reactant through 97 tubes, all of 0.137 cm inside diameter. It is the same reactor described previously in Chapts. II and III. The rate of turbulent energy dissipation ε and the estimated segregation scale L_s for Vassilatos' reactor are shown plotted as a function of distance from reactor entrance in Fig. 19. The values used were obtained

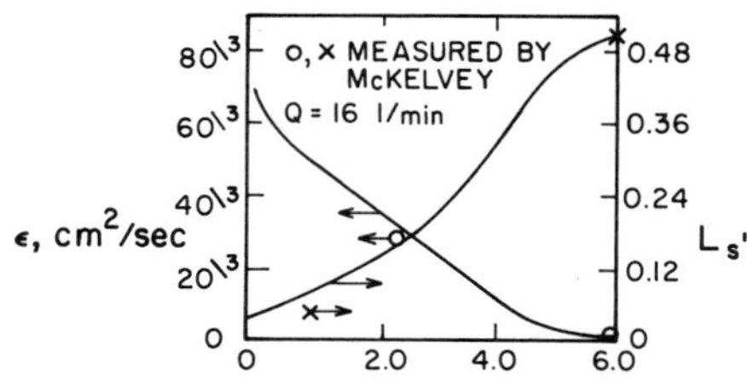

Fig. 19. Turbulent energy dissipation rate and segregation scale for Vassilatos reactor.

by approximating the change of each with distance by use of the points measured by McKelvey (1968) and McKelvey *et al.* (1974). The calculated values of segregation are compared to experimental values in Fig. 20. It is interesting to note

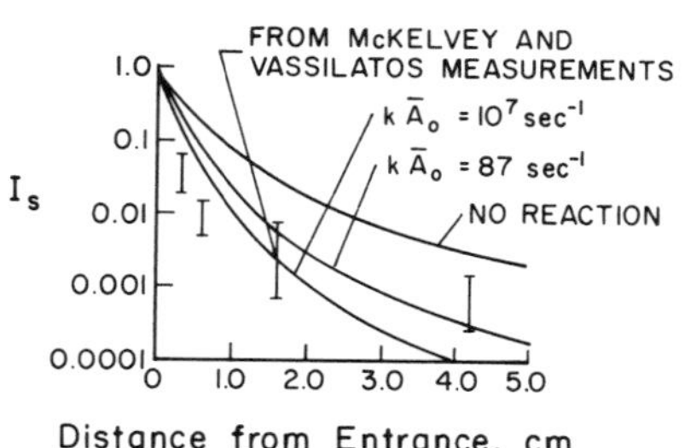

Fig. 20. Comparison of calculated segregation with measured values.

that correspondence occurs only with fast reaction, the effect of γ (see Eq. 6) being to lower segregation values significantly. Such an effect of reaction on segregation has usually been assumed insignificant and has not yet been experimentally investigated. One might be tempted to compare this with the statistical independence hypothesis (see Chapts. I and IV), but this only applies to the premixed case, not to the reactant mixing being considered here.

With the specified hydrodynamic conditions, comparisons of calculated and measured reaction conversion levels as a function of distance from the reactor entrance were made for various reaction rates and reactant ratios as shown in Figs. 21 - 23 (Patterson, 1973). The most significant discrepancy is for the very fast reaction data ($k = 10^{11}$), but this occurs only in the entrance region. The comparisons for the slow reaction data in Fig. 23 show reasonable correspondence, but it should be noted that the

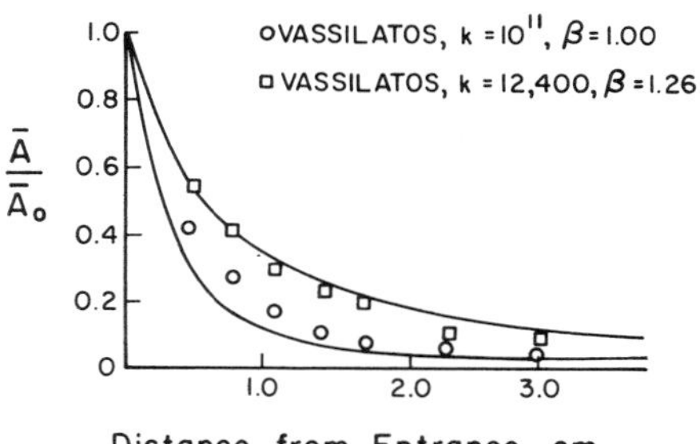

Fig. 21. Comparison of calculated results with Vassilatos' data.

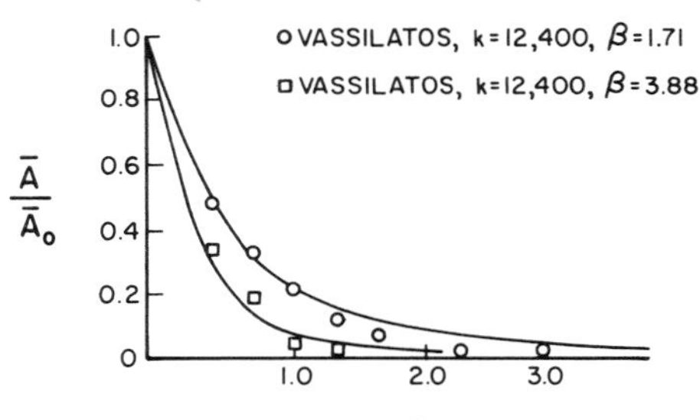

Fig. 22. Comparison of calculated results with Vassilatos' data for high reactant ratios.

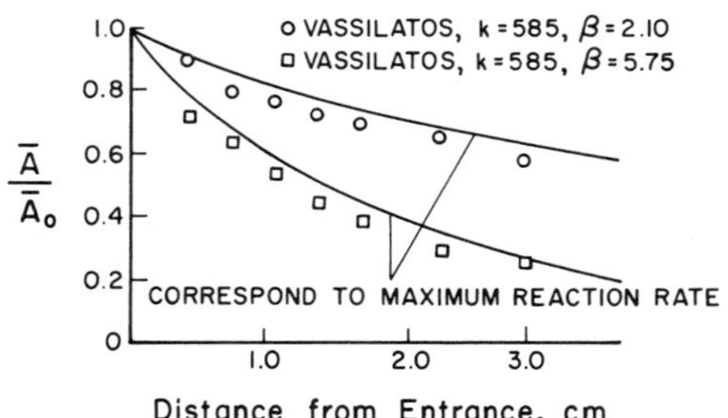

Fig. 23. Comparison of calculated results with Vassilatos' data for slow reactions.

calculated values correspond to maximum possible conversion with plug flow since segregation was much lower than the product of the reactant concentrations. It should be emphasized that the model uses ε and L_s data (which should fit I_s). In contrast, Toor in Chapt III and McKelvey *et al.* (1974) use I_s itself directly from the measured data.

XI. USE OF THE MODEL FOR UNSTEADY RESPONSES TO UPSETS

Modeling of the unsteady response of reactors with second-order reaction and non-premixed reactants has until recently not been attempted. The hdm model should allow such a calculation, and such results have been reported in detail by Otte (1974). It was impossible to find experimental dynamic data for such a system, but the calculated responses to sudden changes of feed rate and impeller speed were reasonable and were affected by parameters such as reaction rate constant in reasonable ways. Without reaction present, it was possible to almost perfectly duplicate the dynamic data of Hubbard and Patel (1971) for a sudden concentration change in the feed.

Of special value is that the response of every segment in the reactor is obtained. In cases where temperatures must not exceed a certain value, for example, startup and rate change upsets may be simulated to test for the occurrence of high temperature transients.

XII. RELATIONSHIP BETWEEN THE MODEL AND RANDOM COALESCENCE-DISPERSION MODELS

The relationship between the dominant mixing parameter in the hdm model, $(\varepsilon/L_s^2)^{1/3}$, and the interaction parameter I in the cd models, average coalescences per site per time increment, can be established. The necessary values of I were those in each reactor segment which would produce the same conversion level as the hdm model. Figure 5 shows such a comparison for $\beta = 1.26$ along with the Vassilatos data. The necessary values of I are plotted against $(\varepsilon/L_s^2)^{1/3}$ in Fig. 24.

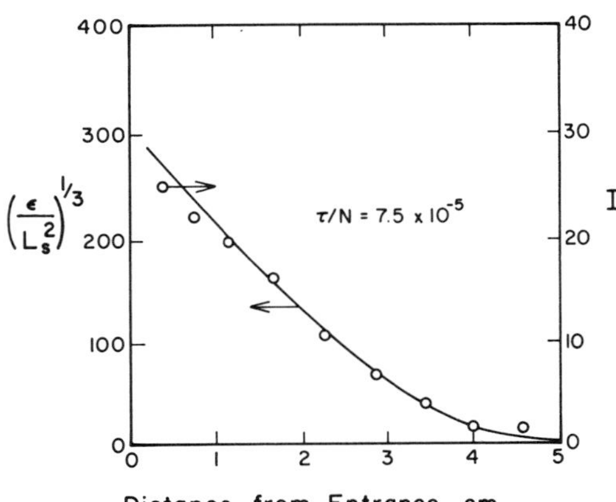

Fig. 24. Comparison of values of HDM parameter $(\varepsilon/L_s^2)^{1/3}$ and of the C-D interaction parameter I for a tubular reactor.

From this, a reasonable fit is

$$I = 1333\ (\overline{\tau}/N)(\varepsilon/L_s^2)^{1/3} \tag{11}$$

where $(\overline{\tau}/N)$ is the residence time per coalescence site. The importance of this relationship (or a better one in the future) is that a knowledge of turbulent energy dissipation rates and length scales in a reactor would enable the use of the far simpler cd model in cases where the kinetics are too complex or money is too limited to use the hdm model. Of course, as pointed out previously, improved estimates for ε and L_s would be helpful, although those given in Chapt. II seem reasonable.

XIII. COMBINED SOLUTION OF THE EQUATIONS OF MOMENTUM AND ENERGY WITH THE EQUATION FOR MASS

A complete solution to the turbulent mixing and diffusion problem must involve a simultaneous solution of the momentum, mass and energy balance equations, if interactions exist, as, for example, when there are buoyancy effects caused by different densities of mixing components. Since many complex terms are involved in these equations, logical ways of modeling each term with as much physical reality as possible are sought. Since the nature of the Reynolds-averaged equations for turbulence (mass, momentum and energy) is to include one more correlation term than available equations, an approximation to the highest order term to allow closure of the set of equations is always necessary. The so-called mean *turbulent closure models* which yield solutions for turbulence quantities, as well as average velocity and shear stress, are the ones which are useful in modeling mixing.

A number of the mean turbulent closure models use the assumption that Reynolds stress is proportional to velocity gradient (Prandtl and Wieghardt, 1945; Harlow and Nakayama, 1967; Nee and Kovasznay, 1968; Ng and Spalding, 1972; and Spalding, 1971). The Harlow-Nakayama and Spalding models involved two undetermined turbulence variables and, hence, two balance equations. However, successful efforts to achieve generality, i.e., to fit a large number of different boundary layers, have resulted from models that use relationships between Reynolds stress and the local turbulent kinetic energy and result in only one turbulence variable to be determined. Bradshaw *et al.* (1967, 1971, 1972), Lee *et al.* (1972) and Bush and Fendell (1973) assumed that Reynolds stress is proportional to turbulence energy as suggested earlier by Kolmogoroff (1942) and Nevzglydov (1945). Mellor and Herring (1973), who reviewed these model closure methods, showed the relationship between them and the more general (complex) models based on the Reynolds-stress models (Rotta, 1951; Donaldson, 1968; Daly and Harlow, 1970; Hanjalic and Launder, 1972; Launder *et al.*, 1973), which account for effects of anisotropy.

Bradshaw (1968) and Spalding (1971) have presented methods for solving the mass balance equation for mixing. The Bradshaw model makes use of an approximation to the term for the $\overline{u_y b}$-correlation which arises in diffusion, but does not yield information on the segregation level (b'). The Spalding model allows prediction both of concentration profiles during diffusion and the segregation of the mixing components. One of the equations recommended involves a balance between scalar intensity convection, diffusion, production and decay. The closure approximation used is that $\overline{u_y b} \propto (\mu_t db/dy)$, where $\mu_t \propto (\rho q'^2/\varepsilon)$. This is equivalent to setting momentum and mass transport proportional to one another, since μ_t is the turbulent viscosity used in the momentum and energy equations. This equation must be solved simultaneously with the hydrodynamic equations if the terms for diffusion are significant compared to the convection terms. The segregation or intensity decay term was modeled by assuming that $-\partial b'^2/\partial t \propto (\varepsilon/L_s^2)^{1/3} b'^2$, which was the dominant part of the successful Corrsin equation. Spalding showed that his formulation, which amounted to a three differential equation model, predicted qualitatively, at least, the proper profiles of b', q', and ε in the fully developed part of a jet that is entraining surrounding fluid. Work like that of Spalding and co-workers has the promises of predicting turbulent diffusion and mixing from a knowledge of only the boundary conditions in many geometries.

The value of these mixing models which include the equations of momentum and energy is in their application to systems where appreciable turbulent diffusion takes place as well as local mixing where the flow patterns are not too complex. It is very difficult, for instance, to use the hdm mixing model for mixing in a pipe or confined jet unless the rate of spread of the injected component is simply conjectured. The simultaneous mixing and diffusion in rivers and in the atmosphere would also require the more complete approach.

XIV. APPLICATION OF AVAILABLE DESIGN AND SCALE-UP METHODS TO INDUSTRIAL PROBLEMS

In general, the presently available design and scale-up methods which utilize rtd and/or two-environment concepts allow determination of the *average* level of conversion in a reactor. The methods most promising for that type of calculation are those listed in Table I. These methods have certain disadvantages, however, when compared to the hdm model or a more complete model which incorporates momentum and turbulence energy equations: 1) They cannot yield a distribution of conversion, segregation and temperature in the reactor. 2) They involve parameters which are difficult to relate to the operating variables of real mixers. 3) In many cases, their application to reactors with non-premixed feeds is impossible or questionable. These one-parameter models are, of course, easy and cheap to use.

The problems for which the hdm model is particularly valuable include the following: 1) Any reactor in which more precise modeling of the *effluent* conversion is necessary using only real parameters. 2) Reactors in which temperature distributions are important. 3) Reactors in which feed momentum and jetting effects are to be investigated. 4) Cases where reaction kinetics are not known, but may be inferred from pilot data by using the hdm model.

Applications which should be made in the near future include extension of the hdm model or variations to competitive and consecutive reactions, auto-catalyzed reactions, and mixing of reactant fluids of greatly differing viscosities. Of course, further investigation of the basic assumptions is necessary. Application of the full turbulence simulation models to mixing in pipes, jets, static turbulent mixers, rivers and the atmosphere requires much further development, but would be possible, at least in principle.

XV. ACKNOWLEDGMENT

Appreciation is expressed for contributions by Leon L. Otte, T. J. Chen, Ronald M. Canon, Kenneth W. Wall and Allen W. Smith to the work described in parts of this paper.

XVI. REFERENCES

Becker, H.A., Rosenweig, R.E. and Gwozdz, J.R. (1966) *A.I.Ch.E.J. 12,* 964.

Bowers, R.H. (1965) *A.I.Ch.E.-I. Chem. E. Symp. Ser. 10,* 14.

Bradshaw, P., Ferriss, D.H. and Atwell, N.P. (1967) *J. Fluid Mech. 28,* 593.

Bradshaw, P. (1968) NPL Aero. Report 1271.

Bradshaw, P. and Ferriss, D.H. (1971) *J. Fluid Mech. 46,* 83.

Bradshaw, P. and Ferriss, D.H. (1972) *J. Basic Eng., Trans. ASME, 94D,* 345.

Brodkey, R.S. (1966) *A.I.Ch.E.J. 12,* 403.

Bush, W.B. and Fendell, F.E. (1973) *Phys. Fluids 16,* 1189.

Canon, R.M., Wall, K.W., Smith, A.W. and Patterson, G.K. (1974) submitted to *Chem. Eng. Sci.*

Chauhan, S.P., Bell, J.P. and Adler, R.J. (1972) *Chem. Eng. Sci. 27,* 585.

Christiansen, D.E. (1969) *Ind. Eng. Chem. Fund. 8,* 263.

Connolly, J.R. and Winter, R.L. (1969) *Chem. Eng. Prog. 65, 8,* 70.

Corrsin, S. (1957) *A.I.Ch.E.J. 3,* 329.

Corrsin, S. (1964) *A.I.Ch.E.J. 10,* 870.

Curl, R.L. (1963) *A.I.Ch.E.J. 9,* 175.

Cutter, L.A. (1966) *A.I.Ch.E.J. 12,* 35.

Daly, B.J. and Harlow, F.H. (1970) *Phys. Fluids 13,* 2634.

Danckwerts, P.V. (1953) *Chem. Eng. Sci. 2,* 1.

Danckwerts, P.V. (1958) *Chem. Eng. Sci. 8,* 93.

Donaldson, C.duP. (1968) AIAA Paper No. 68-38.

Evangelista, J.J. (1970) Ph.D. Thesis, The City College of New York.

Felder, R.M. and Hill, F.B. (1969) *Chem. Eng. Sci. 24,* 385.

Fox. E.A. and Gex, V.E. (1956) *A.I.Ch.E.J. 2,* 539.

Gegner, J.P. and Brodkey, R.S. (1966) *A.I.Ch.E.J. 12,* 817.

Hanjalic, K. and Launder, B.E. (1972) *J. Fluid Mech. 52,* 609.

Harlow, F.H. and Nakayama, P.I. (1967) *Phys. Fluids 10,* 2323.

Hubbard, D.W. and Patel, H. (1969) Preprint 49i, 62nd Ann. Meet. A.I.Ch.E., Wash. D.C.

Kattan, A. and Adler, R.J. (1967) *A.I.Ch.E.J. 13,* 580.

Keairns, D.L. (1969) *Can. J. Chem. Eng. 47,* 395.

Keey, R.B. (1967) *British Chem. Eng. 12,* 1081.

Kim, W.J. and Manning, F.S. (1964) *A.I.Ch.E.J. 10,* 747.

Kolmogoroff, A.N. (1942) *Izv. Akad. Nank. SSSR Ser. Phys. 6,* 56.

Launder, B.E., Reece, G.R. and Rodi, W. (1973) Imperial College, Mech. Eng. Dept., London, Report HTS/73/31.

Lee, J. and Brodkey, R.S. (1964A) *A.I.Ch.E.J. 10,* 187.

Lee, J. and Brodkey, R.S. (1964B) *Rev. Sci. Instr. 34,* 1086.

Lee, S.C., Harsha, P.T., Auiler, J.E. and Lin, C.L. (1972) PROC. 1972 HEAT TRANS. AND FLUID MECH. INST., R. B. Landis and G.J. Hordemann, Eds., Stanford Univ. Press.

Leitman, R.H. (1970) Ph.D. Thesis, Polytechnic Institute of Brooklyn.

Levenspiel, O. (1972) CHEMICAL REACTION ENGINEERING, John Wiley and Sons, New York.

Manning, F.S. and Wilhelm, R.H. (1963) *A.I.Ch.E.J. 9,* 12.

Mao, K.W. and Toor, H.L. (1970) *A.I.Ch.E.J. 16,* 49.

Marr, G.R. and Johnson, E.F. (1963) *A.I.Ch.E.J. 9,* 383.

McKelvey, K.N. (1968) Ph.D. Thesis, The Ohio State University.

McKelvey, K.N., Yieh, H.N., Zakanycz, S. and Brodkey, R.S. (1974) submitted to *A.I.Ch.E.J.*

Mellor, G.L. and Herring, H.J. (1973) *AIAA J. 11,* 590.

Methot, J.C. and Roy, P.H. (1971) *Chem. Eng. Sci. 26,* 569.

Methot, J.C. and Roy, P.H. (1973) *Chem. Eng. Sci. 28,* 1961.

Nee, V.W. and Kovasnay, L.S.G. (1968) Proc. AFOSR-IFP Stanford Conf.

Nevzglydov, V. (1945) *J. of Phys. (USSR) 9,* 235.

Ng, D.Y.C. and Rippin, D.W.T. (1965) THIRD EUROPEAN SYMPOSIUM ON CHEMICAL REACTION ENGINEERING, Amsterdam, Pergamon Press, Oxford, 161.

Ng, K.H. and Spalding, D.B. (1972) *Phys. Fluids 15,* 20.

Nishimura, Y. and Matsubara, M. (1970) *Chem. Eng. Sci. 25,* 1785.

Norwood, K.W. and Metzner, A.B. (1960) *A.I.Ch.E.J. 6,* 432.

Nye, J.O. and Brodkey, R.S. (1967) *Rev. Sci. Instr. 38,* 26.

Otte, L.L., Patterson, G.K. and Chen, T.J. (Nov. 1971) Paper 70d, presented at 64th Annual A.I.Ch.E. Meeting, San Francisco.

Otte, L.L. (1974) Ph.D. Thesis, University of Missouri-Rolla, Rolla, Missouri.

Patterson, G.K. (August 1970) Paper 5.10, presented at Chemeca '70, Melbourne, Australia.

Patterson, G.K. (1973) FLUID MECH. OF MIXING, ASME, Fluids in Eng. Div.

Penny, W.R. (March 1971) *Chem. Engring. 78,* No. 6, 86.

Porcelli, J.V. and Marr, G.R. (1962) *Ind. Eng. Chem. Fund. 1,* 172.

Prandtl, L. and Wieghardt, K. (1945) *Nachr. der Akadr der Wiss. in Gottingen: Math-Phys. Klasse,* p. 6.

Prochazka, J. and Landau, J. (1961) *Coll. Czech. Chem. Commun. 26,* 2961.

Rao, D.P. and Dunn, I.J. (1970) *Chem. Eng. Sci. 25,* 1275.

Rao, D.P. and Edwards, L.L. (1971) *Ind. Eng. Chem. Fund. 10,* 398.

Rao, D.P. and Edwards, L.L. (1973) *Chem. Eng. Sci. 28,* 1179.

Rao, M.A. (1969) Ph.D. Thesis, The Ohio State University.

Rao, M.A. and Brodkey, R.S. (1972) *Chem. Eng. Sci. 27,* 137.

Reith, I.T. (1964) *A.I.Ch.E.-I. Chem. E. Symp. Ser. 10,* 125.

Rippin, D.W.T. (1967) *Chem. Eng. Sci. 22,* 247.

Rosensweig, R.E. (1964) *A.I.Ch.E.J. 10,* 91.

Rosensweig, R.E. (1966) *Can. J. Chem. Eng. 44,* 255.

Rotta, J.C. (1951) *Z. fur Phys. 129,* 547 and *131,* 51.

Rushton, J.H., Costich, E.W. and Everett, J.H. (1950) *Chem. Eng. Prog. 46,* Pt. I, 395 and Pt. II, 467.

Rushton, J.H. (1952) *Chem. Eng. Prog. 48,* 95.

Rushton, J. and Oldshue, J.Y. (1953A) *Chem. Eng. Prog. 49,* 161.

Rushton, J.H. and Oldshue, J.Y. (1953B) *Chem. Eng. Prog. 49,* 267.

Smith, J.M. (1970) CHEMICAL ENGINEERING KINETICS, McGraw-Hill, New York.

Spalding, D.B. (1971) *Chem. Eng. Sci. 26,* 95.

Spielman, L.A. and Levenspiel, O. (1965) *Chem. Eng. Sci. 20,* 247.

Takamatsu, T., Sawada, T. and Izumozaki, N. (1971) *J. Chem. Eng., Japan 4,* 81.

Toor, H.L. (1962) *A.I.Ch.E.J. 8,* 70.

Toor, H.L. (1969) *Ind. Eng. Chem. Fund. 8,* 655.

Treleavin, C.R. and Tobgy, A.H. (1971) *Chem. Eng. Sci. 26,* 1259.

Treleavin, C.R. and Tobgy, A.H. (1972) *Chem. Eng. Sci. 27,* 1653.

Treleavin, C.R. and Tobgy, A.H. (1973) *Chem. Eng. Sci. 28,* 413.

Uhl, V.W. and Gray, J.B. (1966) MIXING THEORY AND PRACTICE, Academic Press, New York.

Vassilatos, G. (1964) Ph.D. Thesis, Carnegie Institute of Technology.

Vassilatos, G. and Toor, H.L. (1965) *A.I.Ch.E.J. 11,* 666.

Villermaux, J. and Zoulalian, A. (1969) *Chem. Eng. Sci. 24,* 1513.

Weinstein, H. and Adler, R.J. (1967) *Chem. Eng. Sci. 22,* 65.

Zeitlin, M.A. and Taylarides, L.L. (1972) PROC. FIFTH EUROPEAN SYMPOSIUM ON CHEMICAL REACTION ENGINEERING, Amsterdam, Holland.

Zwietering, Th.N. (1959) *Chem. Eng. Sci. 11,* 10.

Industrial Turbulent Mixing

L. L. Simpson

Chapter VI

Industrial Turbulent Mixing

L. L. SIMPSON

Engineering Department
Union Carbide Corporation
South Charleston, West Virginia 25303

I. INTRODUCTION

Most industrially important chemical reactions take place because of turbulent mixing. Turbulence forms very small contiguous masses of reactant species, thereby effecting an enormous reduction in the molecular mixing time. Without such phenomena, the cost of many chemicals would be orders of magnitude more than they are today. Turbulent mixing also plays an important role in several other chemical operations including blending, solids suspension, gas-liquid contacting and liquid-liquid contacting. In addition, atmospheric diffusion disperses accidental chemical spills because of turbulence.

This chapter contains design method, evaluation procedures and scale-up techniques that can be applied to industrial turbulent mixing problems. It includes material on pipeline mixing, jet mixing and tubular reactor stratification, together with brief discussions on atmospheric diffusion and bubble and drop size estimation.

The industrial mixing specialist has a bewildering array of design tools at his disposal. They vary from simplistic rules of thumb to sophisticated turbulence theory and from vendor literature to experimental data. He must decide which methods are applicable to his specific problem.

Frequently, he does not have time to thoroughly research a problem and must rely on his intuitive judgement for order of magnitude estimates. Even when he has sufficient time, he often cannot find literature references to the problem that confronts him; consequently, he must turn elsewhere for a solution.

Although turbulence theory has often been dismissed as being too theoretical or impractical, it can be quite useful. The designer might compare mixer designs in terms of the expected eddy sizes or rates of energy dissipation, or he might use it to scale up equipment which is performing satisfactorily.

II. PIPELINE MIXING

II.A. *FUNDAMENTALS*

Mixing is usually accompanied by an expenditure of energy. With miscible fluid mixing, the longer the time available to mix, the lower the required energy input. In theory, miscible fluids could be mixed solely by molecular diffusion; however, the process is usually so slow that the mixing container would have to be intolerably large and expensive. Fortunately, most industrial systems operate in a turbulent regime where mixing can be achieved with a relatively low energy input. In operations such as solids suspension and liquid-liquid dispersion, minimum energy levels are needed to avoid phase separation.

Drawing on the work of Kolmogorov, Landau and Lifshitz (1959) eloquently summarized the role of energy in turbulent mixing. They pointed out that there is a distribution of different eddy sizes in turbulent flows. Eddy size is defined as the distance over which there is an appreciable change in velocity. The largest eddies are the same order of magnitude in size as the apparatus carrying the flow, and they contain most of the kinetic energy. Energy in the turbulence continually cascades from the largest to the smallest eddies where it dissipates as heat. The smallest eddies have a Reynolds number of the order of one and a size of the order of

$$\lambda_0 \simeq (\nu^3/\varepsilon)^{1/4} \tag{1}$$

where ν is the kinematic viscosity and ε is the rate of energy dissipation per unit mass. In a mixing vessel, ε can be calculated from the horsepower and liquid mass. An energy input of 1 hp/1000 gal in a water filled vessel is equivalent to $\varepsilon = 0.20\ m^2/s^3$. In a six inch pipe, a water velocity of 6 fps gives an average $\varepsilon = 0.17\ m^2/s^3$, while 100 fps of atmospheric air gives $\varepsilon = 780\ m^2/s^3$.

At the λ_0 scale, mixing occurs by a complex interaction between turbulence and molecular diffusion (Brodkey, 1967). It is easily seen from Eq. (1) that higher rates of energy dissipation will reduce λ_0, resulting in more rapid mixing. From this idea, it is also possible to show that mixing is favored by high velocities in small apparatus. Although Kolmogorov's theory does not provide a design method, it does show the importance of equipment size and energy dissipation in design. It has also paved the way for more sophisticated theories that are of value to the designer.

Modern turbulence theory has emphasized the use of root-mean-squared (rms) concentration fluctuations instead of time averaged concentrations. The distinction is important. An analysis technique that measures a time averaged concentration cannot detect the rapid concentration variations present in all turbulent mixing processes; hence, it cannot determine when true molecular mixing has been achieved.

II.B. *PIPING AND TEES*

In existing chemical processing units, it is common to find that two fluids have been combined at a tee. Often, no special precautions have been taken to ensure high quality mixing; yet, more often than not, mixing quality is acceptable. A problem can arise, however, when attempts are made to upgrade the operation. Mixing quality becomes suspect and meetings are held to decide how to improve it or, indeed, whether it should be improved. In such circumstances, one needs to be able to predict mixing quality in piping.

Chilton and Genereaux (1930) undertook one of the first pipe mixing studies. They visually observed mixing of smoke-traced secondary gases (air, sulfur dioxide, carbon dioxide and ammonia) injected into glass tee branches. The mainstream air flowed through a 1.75 inch diameter line portion of the tee. The branch diameter was varied between 0.25 and 1.5 inches. They observed that good mixing was achieved in two to three pipe diameters if the tee branch gas was injected at a mass velocity two to three times that of the primary gas entering the tee. Smaller mass velocities prevented the injected gas from penetrating into the mainstreams, and higher ratios caused the injected gas to overpenetrate. The mass velocity needed for good mixing decreased with increasing branch size. Although Chilton and Genereaux tried other injection methods into the 1.75 inch pipe, none exhibited a marked advantage over the simple tee. Chilton and Genereaux probably would have been more successful if they had correlated the ratio of the velocity times the square root of density for the two fluids; however, their experimental technique probably did not justify added sophistication. The visual smoke tracer method is not a good mixing quality measure. Aside from its obvious quantitative deficiency, the method cannot detect small scale mixing. Nor will it detect rapid concentration fluctuations.

The recent tee-mixing study of Narayan (1971) was more quantitative than the earlier work of Chilton and Genereaux. Narayan mixed air with an air-carbon dioxide mixture and used the specific gravity of the sampled stream at various stages of mixedness to measure mixing quality. His mixers are shown schematically in Fig. 1. With his types II

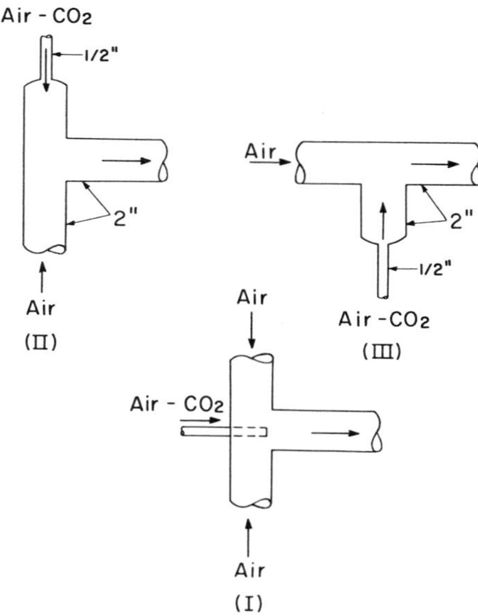

Fig. 1. Mixing devices tested by Narayan (1971).

and III mixers, he injected the unmixed gases perpendicularly to each other. Reynolds numbers were roughly 50,000. As indicated by the specific gravity difference between the center and periphery samples, he achieved high quality mixing at the first downstream sample point, located 4.4 and 8.7 diameters downstream of the types II and III mixers, respectively. The ratio of the velocity in the 1/2 inch line to that in the 2 inch line was approximately 3.5. Narayan's tee results agree with those of Chilton and Genereaux; both indicate that rapid gas mixing can be realized with a simple tee.

Narayan also studied pipeline mixing with the type I mixer. Whereas he had been able to achieve quality mixing in a few diameters with the perpendicular flow mixers, this parallel flow device required 250 diameters. This result is an important one, because it gives the designer an indication of how bad pipeline mixing can be, and it illustrates the need for pipeline mixing technology.

Parallel-flow mixing was also the subject of studies by Hiby (1971) and by Hartung and Hiby (1972). Both studies used 19 mm inside diameter pipe. Introduced at equal velocities into the pipe, acid and base solutions were separated by a thin diaphragm. Mixing began at the end of the diaphragm. They used colorimetric methods to measure a mixing quality indicator, termed the intensity of segregation (not to be confused with intensity of segregation based on rms concentration fluctuations), which is defined as

$$S = (\overline{A} - \overline{A}_0)^2 / \overline{A}_0(1 - \overline{A}_0) \qquad (2)$$

where $\overline{A}$ and $\overline{A}_0$ are the fractional base concentration and the initial fraction of base solution introduced into the pipe. Since $\overline{A}_0 = 1/2$ in the tests, $\sqrt{S} = \sigma_a / \overline{A}_0$ where σ_a is the standard deviation. With $\sigma_a / \overline{A}_0 = 0.01$, the mixing length-to-diameter ratio is 90 at a Reynolds number of 6,000. Other Reynolds numbers in the 3,000 to 17,000 range gave somewhat larger mixing lengths. Hiby noted that the presence of a bend would reduce the mixing length.

II.C. *TURBULENCE ANALYSIS*

During the 1950's, a few researchers began applying turbulence theory to mixing phenomena. The results of their work were of little value to the industrial mixing specialist who needed solutions to his practical problems. Turbulence theory, however, cannot be judged on the basis of its direct applicability to practical problems. Its real value is a result of its being a theoretical approach. It provides the specialist with an insight into mixing that he cannot achieve either experimentally or through the use of other technology.

Hughes (1957) combined Einstein's diffusion equation with Kolmogorov's postulate to predict minimum mixing times. He showed that the minimum mixing time for a liquid is substantially greater than that for a gas, because of the lower molecular diffusion coefficient in the liquid. It is only with very fast reactions that gaseous diffusion effects are likely to be important.

Using turbulence theory, Beek and Miller (1959) estimated gas and liquid pipeline mixing lengths as a function of both the Reynolds number (N_{Re}) and the number of injection nozzles. With one nozzle they found that 30 to 40 pipe diameters (depending on N_{Re}) were needed to reduce the rms concentration fluctuation (a') of a gas to ten percent of its initial value. If 64 injection nozzles were used instead of one, they estimated that the mixing length would be reduced to roughly six diameters. With single nozzle injection of a liquid ($N_{Sc} = 1{,}000$), the mixing length-to-diameter ratio varied from 120 at a N_{Re} of 25,000 to 67 at a N_{Re} of 100,000. Multiple nozzle injection with liquids resulted in only minor reduction in mixing length. Even though the Beek and Miller mixing lengths are roughly correct, their findings for liquids have not been verified. A correlation obtained by Hartung and Hiby (1971), for example, showed that liquid mixing length increased with Reynolds number for N_{Re} above 6,000. Their theoretical finding concerning the benefits of multiple injection nozzles is also suspect. Recently, Singh and Toor (1974) found that multiple injection mixing was substantially faster than single jet mixing.

Lee and Brodkey (1964), Brodkey (1966) and Gegner and Brodkey (1966) applied the isotropic turbulent mixing theory of Corrsin (1957, 1964) to pipeline mixing and showed that it was valid over the central portion of a pipe. Because his work has led to some practical results, the theory will be reviewed here; more details can be found in Chapt. II.

With stationary isotropic turbulence, the square of the rms concentration fluctuation will decay exponentially from an initial value. Mathematically,

$$I_s = (a'/a'_0)^2 = e^{-t/\tau} \tag{3}$$

where t is the time and τ is a time constant. Time is taken equal to the distance from a tracer injection point divided by the time average velocity at the point of interest in the pipe, or

$$t = z/\overline{U}_z \tag{4}$$

The time constant was given by Corrsin as a function of the Schmidt number:

$$\tau = (5/\pi)^{2/3}(L_s^2/\varepsilon)^{1/3}[2/(3 - N_{Sc}^2)] \text{ for } N_{Sc} < 1 \tag{5}$$

$$\tau = (1/2)[3(5/\pi)^{2/3}(L_s^2/\varepsilon)^{1/3} + (\nu/\varepsilon)^{1/2}\ln N_{Sc}] \text{ for } N_{Sc} >> 1 \tag{6}$$

N_{Sc} is the Schmidt number and L_s is a scalar integral length scale. Brodkey (1966) and Gegner and Brodkey (1966) reported the following correlations of the parameters:

$$(5/\pi)^{2/3}(L_s/\varepsilon)^{1/3} = 0.341r_0/u_z' \qquad (7)$$

$$\varepsilon = 4.4u_z'^3/r_0 \qquad (8)$$

where r_0 and u_z' are the pipe radius and rms axial velocity fluctuation; u_z' can be estimated from Laufer's results (1954). Eq. (8) is in good agreement with Laufer's findings at the centerline, but it overestimates the dissipation at most other locations in the pipe.

This analysis indicates that the mixing length-to-diameter ratio is a function of the Schmidt number, Reynolds number and the concentration criterion. Results for $a'/a_0' =$ .01 are shown in Fig. 2. This graph represents an extrapolation of the Reynolds number, Schmidt number and concentration criterion.

The dispersed plug-flow model given by Levenspiel and Bischoff (1963) was used as a further check on pipe mixing lengths. Their radial dispersion coefficients, together with the axial coefficients suggested by Woodhead *et al.* (1971), were used in the model. Mixing length was defined as the dimensionless length required for a centerline tracer concentration to drop to 1.01 times its equilibrium value. Results given in Fig. 2 are similar to the turbulence results, even though the calculations and concentration criteria are totally different.

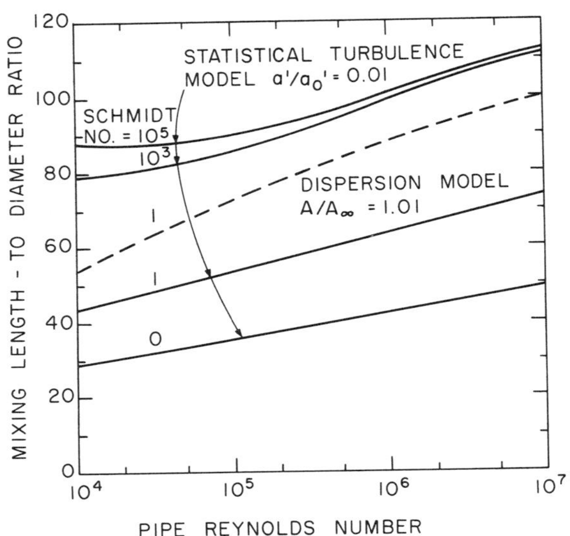

Fig. 2. Pipe mixing length (after Simpson, 1974)

II.D. *FLOW RATE CONTROL*

The importance of good control in reactant pipeline mixing cannot be overemphasized. There is usually little backmixing in pipeline mixers; consequently, even minor flow oscillations will be transmitted downstream and have a very detrimental effect on a reactor.

Pressure drop taken to inject one reactant into another improves control by dampening flow disturbances. Pressure drop taken after the reactants are combined is detrimental because a minor flow rate rise in one of the streams will cause the other stream to decrease, thereby changing the flow ratio and magnifying the disturbance.

The effect of backmixing can be illustrated with an example. Assume that a slug of one liquid, B diameters in length, is initially trapped in a long length of piping. With an axial dispersion calculation for high Reynolds number, it can be shown that that slug would have to move downstream about 5300 B^2 diameters before its concentration would be reduced to one percent of its initial value.

Based on a simplified model of tee mixing, Narayan (1971) concluded that small pressure fluctuations occurring in two different gas feed streams would lead to large composition changes in the combined stream. Reed (1974) noted that pressure variations of one percent or more, having a period of one to two seconds, may exist in a gas system even when a conventional pressure instrument indicates no pressure variation. He speculates that the cyclic changes are caused by the great compressibility of the gas upstream and downstream of the control device.

II.E. *INDUSTRIAL MIXER DESIGN*

Anyone who attempts to specify an in-line mixer is confronted with a bewildering number of alternatives. Even if he decides to buy a commercially available unit, there are several acceptable alternatives. The problem is worse if he designs it himself. Should he use internal baffles in the pipe? Perhaps an internal sparger would be more suitable.

There are several reasons for choosing an in-house design over a commercial design. One of the fluids to be mixed could be highly reactive and form undesirable products if not diluted very rapidly. Fluid compositions might, for example, pass through the explosive range. A second reason for designing the mixers is that commercially available units may not fit the process requirements. The pressure drop available could be too small, as is often the case with large gas flows, or materials and mechanical limitations might prevent the use of commercial mixers. Finally, it might be less expensive to design a mixer in-house.

As a first step in the design process, the designer should decide what he is trying to accomplish with the device and what are the design limitations. Is mixing time important? If so, how quickly must the two fluids be mixed? What is the pressure drop limitation? Is turndown required? Are there any limitations caused by the materials of construction or by the design pressure?

In discriminating between alternatives, the designer should place great emphasis on performance predictability. The mixer must behave as he predicts it will. It is particularly important that he be able to predict pressure drop. Too often, mixers are installed and later must be removed after causing excessive pressure drop. This problem will not occur as frequently if only about half of the available pressure drop is used for design.

Mixers must be designed for mechanical integrity. Distribution piping installed in a crossflowing stream can fail from flow induced vibration. Baffle plates must withstand the pressure drop across them.

Probably the most important functional considerations in mixing effectiveness are the mixing scale and the rate of energy dissipation per unit mass. As indicated by Eqs. (5) and (6), these two factors determine the mixing time constant. The specific geometric designs considered in this section can be evaluated in terms of mixing scale and energy dissipation rate. Low energy dissipation mixers usually require comparatively long residence times for a given mixing quality; nevertheless, such mixers may be attractive because of their low pressure drop. For this reason, an open pipe is a highly efficient mixer. Whereas the need for high energy dissipation rates depends on the design objective, all mixers benefit from scale reduction.

Four in-house mixer designs are depicted in Fig. 3

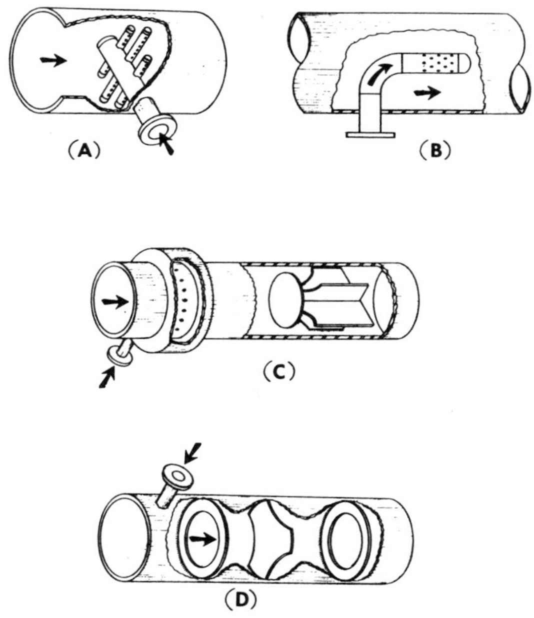

Fig. 3. Industrial mixer designs (after Simpson, 1974).

(Simpson, 1974). The type A mixer has been used to achieve very rapid low-pressure-drop mixing in comparatively large (e.g., 18") gas lines. Each branch pipe or finger has a large number of small orifices drilled perpendicular to the direction of flow. To minimize interference between adjacent orifices, the orifice pitch should be a minimum of about three orifice diameters. The following design steps should lead to a sound design of the type A mixer: 1) Determine the design pressure drop from that available. 2) Taking the orifice pressure drop as three velocity heads, estimate the orifice velocity from the design pressure drop. The velocity should be substantially greater than the crossflowing velocity. 3) From the velocity and the design volumetric flow rate, calculate the total orifice area. 4) Assume that 50 to 200 orifices will be drilled, and calculate an orifice size. 5) Use the closest standard drill size, and re-estimate the number of orifices. 6) With an orifice pitch-to-diameter ratio of three, estimate the number of fingers needed. If there are too many required, re-estimate the orifice size on the basis of a practical number of fingers. The orifice size will be approximately proportional to the number of fingers. 7) Calculate the orifice Reynolds number. It should be above 5,000. 8) Good flow distribution along the length of each finger will be assured if the finger cross sectional area is a minimum of 1.5 times the total orifice area in that finger. Much lower area ratios will cause most of the fluid to flow out the end of each finger. Larger ratios will contribute to unnecessarily higher pressure drop in the main pipe. The area of large inlet distribution pipe should be roughly equal to the total cross sectional area of all the fingers. 9) To estimate the pressure drop through the main pipe, first calculate the flow Reynolds number based on finger diameter and the main velocity (U_m) through the minimum free area of the mixer. Next, determine the drag coefficient from a standard plot of Reynolds number versus drag coefficient. Finally, using U_m and the drag coefficient, calculate the drag force and divide it by the total cross sectional area of the main pipe. The result will be an estimate of the pressure drop. This procedure was used by Boucher and Lapple (1948) to estimate the pressure drop across a single row of tubes. If there is significant mass flow rate of injected fluid, when compared with the mainstream flow rate, an additional pressure drop due to momentum changes should be added to the drag type pressure drop. The momentum contribution to pressure drop is roughly equal to the change in the mass flow rate velocity product around the mixer, divided by the pipe area. 10) Mixing time and jet penetration can be checked with the methods given in a later section. 11) Calculate the vortex shedding frequency from $n_f d_f / U_m = 0.2$ where n_f is the frequency and d_f is the finger diameter. If n_f is fifty percent or more of the natural frequency of a finger, the finger should be supported at its end. If there is any doubt as to whether or not a finger will vibrate, additional support should be provided because vortex shedding is not totally predictable. 12) Multiple rows of holes on one side of a finger should be avoided. If the crossflow velocity head is as much as ten percent of the orifice pressure drop, high circumferential flow maldistribution can result from multiple rows of holes because of the pressure variation around a given finger. See Schlicting (1968). 13) There should be sufficient lengths of main piping upstream and downstream of the mixer to avoid pressure and velocity variations across the main pipe. Ito (1960) and Ward-Smith (1971) present data on the pressure distribution around an elbow. 14) Because of the cost of energy, the uncertainty associated with the crossflow and orifice pressure drops and the velocity and pressure maldistribution suggested in item 13, the main piping upstream and downstream of the mixer should be adequately sized. If the designer is uncertain as to which of two sizes to use, he should select the larger one.

Except for Step 12, these design methods are applicable to the type B design, since it is similar to the type A design. Multiple nozzle entries with the type B mixer can be used to reduce the mixing scale. The type B mixer has been installed in both gas and liquid services.

The type C mixer has been used primarily for rapid mixing of miscible liquid streams, and for rapid dilution of a reactant for a fast gas phase reaction. It is particularly suitable in situations where the injected stream would flash if not diluted very rapidly. After the injected stream is diluted, it is combined with the unmixed main-steam fluid through the downstream annular orifice. Typically, both the injection orifice and annular orifice would have a few psi pressure drop across them.

Baffle mixers similar to the type D mixer are in common use. They are easy to design and can be installed in existing piping systems by oversizing the upstream orifice baffle and placing it between flanges. Three tie rods and associated spacers will secure the remaining downstream baffles. Typically, three or five baffles are used and they are placed roughly two pipe diameters apart. Each fractional orifice or hole orifice in the three orifice baffle has its center contiguous with the inside of the pipe. The radius r_h (shown in Fig. 4) is adjusted to achieve the design pressure drop. The design procedure includes the following steps: 1) Select a design pressure drop, divide equally among the baffles, and calculate the associated number of velocity heads N_{VH} per baffle based on the superficial pipe velocity. 2) Refer to Fig. 4 and use Table I to calculate the fractional open area α, orifice radius r_i, hole radius r_h and the minimum web dimension b (the α-N_{VH} relationship is taken from Simpson, 1968). 3) Check the mechanical integrity of the

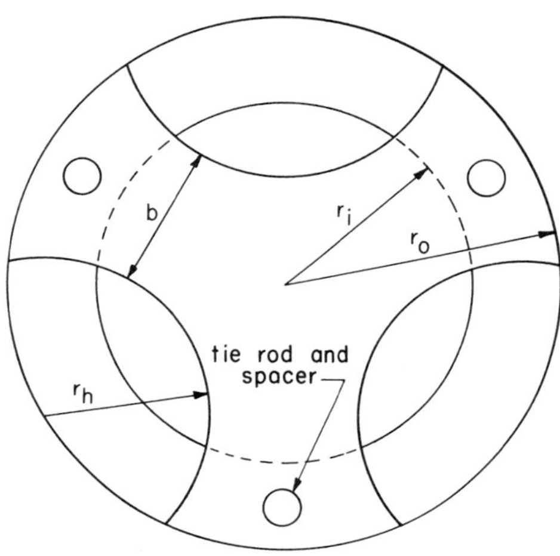

Fig. 4. Type D mixer configuration.

TABLE 1 *Type D Mixer Design Parameters*

N_{VH}	α	r_i/r_0	r_h/r_0	b/r_0
.5	.7583	.8708	.7796	.1729
1	.6847	.8275	.7365	.2590
2	.5994	.7742	.6845	.3631
5	.4762	.6901	.6039	.5244
10	.3837	.6194	.5376	.6569
20	.2990	.5468	.4707	.7907
50	.2063	.4542	.3871	.9578
100	.1523	.3902	.3304	1.0712
200	.1109	.3330	.2804	1.1713
500	.0719	.2682	.2244	1.2834
1000	.0515	.2269	.1891	1.3539

web using the plate thickness and web dimension.

Corrosion, erosion, fouling, construction costs and the mixing objective should all be considered when selecting an orifice size for the types A, B and C mixers. Small orifices subjected to a corrosive environment will undergo a larger fractional change in area than will large orifices during the same period. If fouling is anticipated, operating personnel may prefer a comparatively large orifice -- perhaps one-half inch diameter or larger. In a clean surface, orifices smaller than 0.02 inches in diameter have been used; however, one eighth inch diameter is a more practical minimum.

Orifice drilling costs can be unnecessarily high if the design incorporates five hundred or more orifices. If a large number is required, the designer should consider the use of commercially available perforated plate.

Rapid dilution of an injected stream is best achieved with high orifice velocities and small orifices. Overall mixing quality is favored by high orifice velocities and large orifices (Walker and Kors, 1973). A later section in this chapter will treat this subject in greater depth.

A Schaschlik baffle mixer was tested by Hartung and Hiby (1972). As shown in Fig. 5, this unit consists of

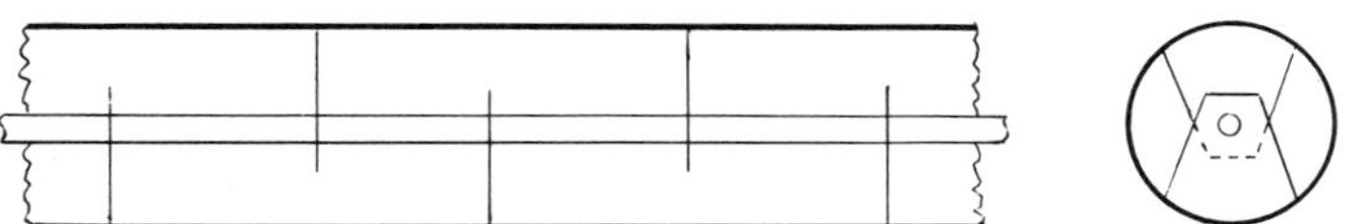

Fig. 5. Schaschlik mixer (after Hartung and Hiby, 1972).

a series of baffles; each one occupies one-half of the pipe cross section and is spaced one pipe diameter from the adjacent baffle. An intensity of segregation of 10^{-4} was achieved in four to five diameters when this unit was operated with the liquids flowing at a Reynolds number of 8,000. Hartung and Hiby data indicate a pressure drop of about seven velocity heads per baffle. If considered as separate orifices, the estimated pressure drop would be about four velocity heads per baffle.

The number of different mixer designs that has been tried is staggering. It is probably at least as great as the number of designers in the field. In addition to the designs already discussed, numerous others are mentioned in literature. Several types are summarized by Treybal (1963). Jacobs (1973) has described a crossflow baffle mixer as well as a tangential entry device. The John Zinc Company is marketing a mixer patented by Reed (1965). It consists of a mixing chamber that contains an annular passage and a central internal pipe. Mixing is achieved by impinging radially outward flowing jets from the central pipe and inward flowing jets from the annular passage. One fluid flows through the pipe orifices and annular jet orifices. Other commercial mixers are discussed in the next section. In a patent assigned to the Shell Oil Company, Son (1972) describes the use of patented internal ring pipes for gas-gas mixing. The injected gas flows through uniformly distributed orifices drilled in each ring at the downstream location. Again, multiple orifices are used to reduce the scale of the mixing.

Gas-gas mixing plays a central role in the design of burners -- particularly those used to achieve high combustion intensities. As a result of their widespread use, different mixer designs abound in the combustion literature (see Beér and Chigier, 1972).

II.F. *COMMERCIAL MOTIONLESS MIXERS*

In recent years, commercially available stationary or motionless mixers have become increasingly popular. Terms such as static and motionless are often used to identify these mixers because they have no moving parts. Frequently developed for mixing of viscous fluids, these devices are also widely used for turbulent mixing. Although their functional superiority in turbulent operations can be questioned, they do present certain advantages over in-house designed mixers: 1) Pressure drop of commercial mixers is predictable. 2) Based on extensive testing, the manufacturer generally understands the performance of his mixer. 3) Engineers who do not have ready access to a specialist can solve mixing problems with a commercial design. 4) The non-specialist may resist the use of a pipeline or tee mixer because of the belief that mixing does not occur easily in piping.

There are some drawbacks. By comparison with the in-house designs, these mixers usually take longer to procure, often have higher pressure drop and may not satisfy the mixing objectives. Moreover, it may be difficult to decide which commercial mixer to use.

Many of the commercially available mixers listed in Table II and shown in Fig. 6 are flow splitting devices. They were conceived with the notion that, as a fluid flows through successive mixing stages, it subdivides into thinner and thinner strata, and finally diffuses into homogeneity. The concept is valid for laminar flow, but it does not depict turbulence accurately. Turbulent flows naturally subdivide without being split. They benefit from large scale radial motion but do not combine by splitting, per se. Fluids in

TABLE II *Motionless Mixer Manufacturers*

MIXER	MANUFACTURER
Durco Mixing Nozzle	The Duriron Company, Inc.
John Zink Inertial Homogenizer	John Zink Company
Kenics Static Mixer^R	Kenics Corporation
Koch Static Mixing Unit	Koch Engineering Company, Inc.
Ross LPD Motionless Mixer	Charles Ross & Son Company
Ross ISG Motionless Mixer	Charles Ross & Son Company

Fig. 6A.

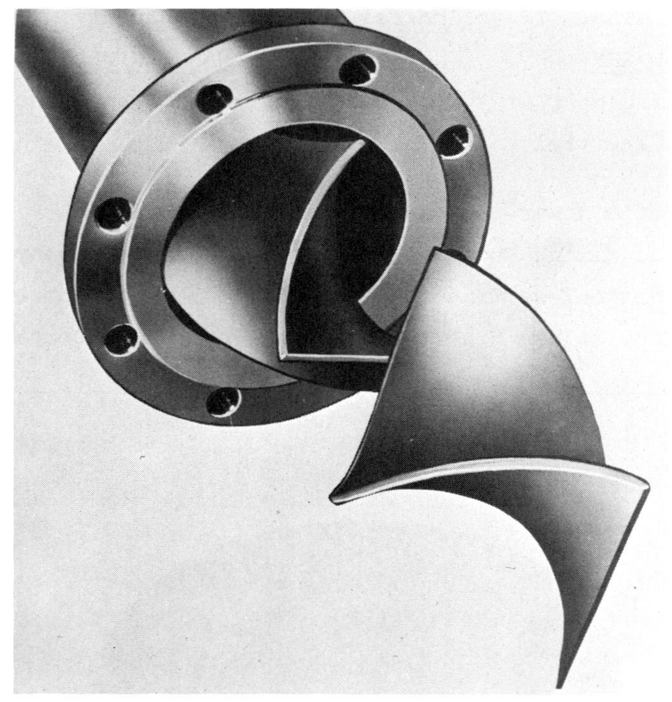

Fig. 6B.

Fig. 6C.

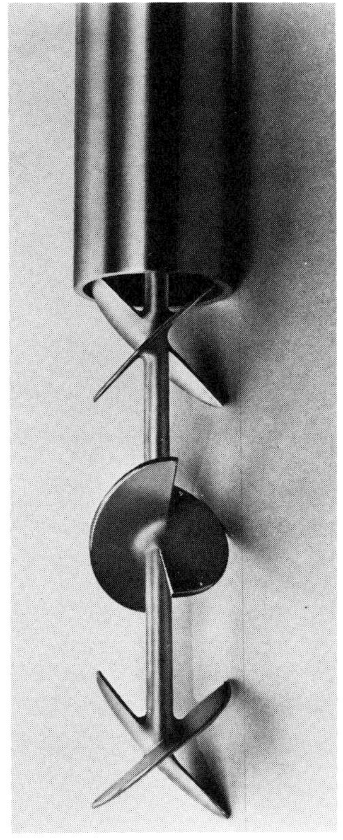

Fig. 6D

Fig. 6. Commercial designs: 6A Durco Mixing Nozzle, 6B Kenics Static Mixer[R], 6C Koch Static Mixing Unit, 6D Ross LPD Motionless Mixer.

turbulent flow can separate from flow splitting surfaces, thereby increasing the system pressure drop without a commensurate inprovement in mixing quality. Flow separation is the major reason for the inefficiency of the type D mixer and of the Schaschlik mixer.

Unless the designer or manufacturer is willing to undertake a test program, the designer usually has to rely on the manufacturer's assessment of mixing quality. For a given project, such tests are seldom justified for turbulent mixing of miscible fluids. Data for one commercial mixer, a Kenics unit, has been reported by Hartung and Hiby (1972). They found that the intensity of segregation decreased to $S = 10^{-4}$ in nine diameters at a Reynolds number of 8,000. For Reynolds numbers between 4,500 and 17,000, they could not establish a correlation between Reynolds number and the intensity of segregation.

Table III summarizes manufacturers' recommendations for liquid mixing, including the number of elements, the

TABLE III *Pressure Drop Through Motionless Mixers**

Manufacturer	No. elements for low viscosity liquid mixing	Approximate length-to-diameter ratio of mixer	Approximate No. of velocity heads pressure drop (N_{VH})
Kenics	6	9	Depends on diameter & Reynolds No. typically 10 to 20
Ross LPD	6	9	45
Ross ISG	4-6	3.6-6.8	360
Koch	3	3	Depends on size of packing typically 10 to 30
Duriron	1	Installed between flanges	Depends on size Range 130 to 230

*Data from manufacturers' literature

length of a mixer and the pressure drop. It was necessary to convert the manufacturers' pressure drop correlations to a common basis. Number of velocity heads was selected because it is relatively insensitive to the Reynolds number and the size in geometrically similar equipment. Because it violates the similarity principle in that different packing sizes can be used in one size of piping, pressure drop in Koch mixers is highly dependent on packing size. When Hartung and Hiby tested the Kenics unit, they measured a pressure drop of only about four velocity heads. This value is probably lower than the manufacturers' data because Hartung and Hiby used a smooth glass tube. In addition, their downstream static pressure could be in error because of the natural vortex produced by the twisted tapes. Wall static pressures are expected to be higher than pipe centerline pressures.

These mixers should not be judged solely on the basis of pressure drop. In some cases, it may be desirable to oversize the line piping to achieve a certain mixing objective. For example, if the objective were to achieve quality mixing with a minimum residence time, while meeting the allowable pressure drop, the designer might argue as follows: Since the pressure drop, volumetric flow rate and fluid density are fixed by the design situation, a function of mixer design parameters is sought that will minimize residence time or void volume of the mixer. The following equations relate to the problem

$$-\Delta p = N_{VH} \rho U_s^2/2 \tag{9}$$

$$V = \alpha L \pi d_m^2/4 \tag{10}$$

$$L = \xi d_m \tag{11}$$

$$Q = U_s^2 \pi d_m^2/4 \tag{12}$$

In these Equations, V, Q, Δp, ρ, U_s, α, L, d_m and ξ are the mixer volume, volumetric flow, pressure drop, fluid density, superficial velocity, mixer void fraction, length, diameter and length-to-diameter ratio, respectively.

From Eqs. (9) - (12), it is easily shown that, for fixed Q and $-\Delta p$, minimizing the function k will minimize the mixer void volume and residence time

$$k = \alpha \xi N_{VH}^{.75} \qquad (13)$$

A four element Ross ISG mixer has a voidage of only about 0.17 so that $k = 0.17 \times 4 \times 360^{0.75} = 56$, a value that compares quite favorably with that of several other commercial mixers. By comparison, the pipeline mixer functionally depicted in Fig. 2 has a mixing length ξ approximately equal to 100 at a Reynolds number of 10^6. With a Fanning friction factor $f = 0.004$, $N_{VH} = 1.6$ and $k = 1.2$. Consequently, the pipeline mixer would have a very low residence time. It cannot be recommended for general use, however, because such a design could require highly erosive velocities in the pipe. With fixed piping size, a pipeline mixer with $N_{VH} = 1.6$ would require much lower pressure drop than all of the commercial mixers. A simple tee mixer would have even lower pressure drop and the volume would be less than that of a pipeline mixer.

Before selecting a mixer, the designer should consider whether or not he has the time and expertise to design a mixer himself. He must also ascertain whether or not there is sufficient project time to purchase a commercial unit. Is rapid mixing needed? Would a tee mixer suffice? How important is pressure drop?

III. JET MIXING

III.A. *JET PHENOMENA*

This section will contain a brief description of selected jet behavior. Several applications will be discussed in later sections.

Fluid jets of all types play a very important role in many turbulent mixing operations, including mixing in piping (e.g., Shell mixer and types A, B, C and D mixers), jet mixing in tanks, combustion mixing and stack exhaust mixing.

Because of their importance, a large number of articles have been written on various aspects of jet behavior. Axisymmetric jets, plane jets, plane wall jets, confined jets, radial wall jets, jets with swirl, jets in a crossflow-all of these types have been the subject of numerous studies. Among the parameters that have been studied are velocity profile, entrainment rate, rms velocity fluctuations, rate of energy dissipation and turbulence scales (Friehe, *et al.* 1972 and Laurence 1956), intermittency factor, mixing time and penetration length. Results from these studies range from detailed turbulence data to simple design correlations.

Costly design errors are often made because design engineers do not recognize the existence of jets, much less understand their behavior. When a fluid flows through an orifice, nozzle or pipeline into a larger chamber, it produces a jet in that chamber. The jet diameter will increase as the distance from the exit increases. Typically, the total included angle of the jet is fourteen degrees. Because of this spreading tendency, surrounding fluid is sucked, or entrained, into the jet to conserve momentum. It is this entraining phenomena that is of importance to the designer.

Jets may be either laminar or turbulent, depending on the Reynolds number. For the common axisymmetric (circular) jets, the characteristic Reynolds number is constant throughout the jet structure and is equal to that at the exit nozzle or orifice. When this Reynolds number is less than roughly 300, the jet will be laminar. Above 300, some disturbances in the flow are amplified (Townsend, 1961). When the Reynolds number is less than roughly 1,000, a developed turbulent flow cannot be sustained. This latter Reynolds number is analogous to the laminar turbulent transition of 2,100 commonly used with pipe flow.

Batchelor (1967) has used the equations of motion to study the behavior of circular laminar jets. The result of this study indicates that for $N_{Re} > 20$, the combined mass flow rate of nozzle fluid plus entrained fluid increases linearly with distance z from the nozzle in accordance with the following equation

$$w_j = 8\pi\mu z \tag{14}$$

where w_j, μ and z are mass flow rate, viscosity and distance, respectively. This result is unexpected because it predicts that the rate of flow in the laminar jet is independent of the flow rate through the nozzle. Apparently, higher nozzle flow rates are counteracted by a narrower jet expansion angle which reduces the entrained flow to maintain a constant flow rate.

Laminar jets are not effective mixing devices. If a laminar jet of viscous liquid were directed into a large tank, the jet would promote movement of the tank liquid; however, the nozzle liquid would not mix with the tank liquid on a molecular scale. As a result, the tank would be found to contain regions of relatively pure nozzle or tank liquid. Effective mixing can take place only through molecular diffusion which occurs at a negligible rate in viscous liquids (unless the nozzle is extremely small).

Some of the features of turbulent jets are illustrated in Fig. 7. In region I, termed the potential core,

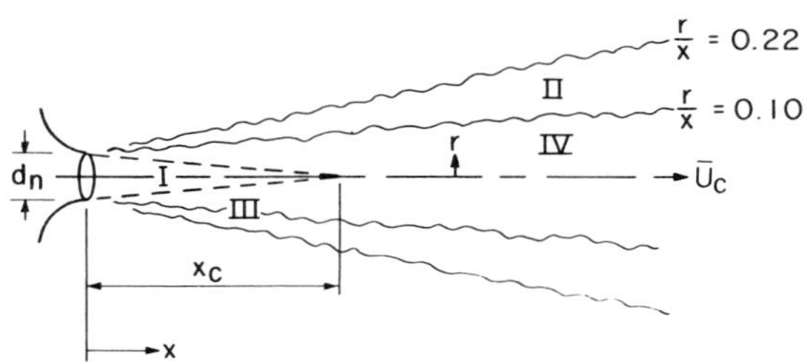

Fig. 7. Axisymmetric free jet, x = z, $x_c = z_c$.

velocity is equal to that in the nozzle itself. The potential core length is designated z_c. Region II is an intermittency zone in which the flow is both turbulent and nonturbulent. Measurements of Corrsin and Kistler (1954) show that for $z/d_n > 10$, flow will be totally turbulent when r/z is less than roughly 0.1. For $0.1 < r/z < .22$, the fraction of the time that turbulence exists, Ω, decreases from near 1 at f/z = 0.1 to near 0 at r/z = 0.22. The parameter Ω is called an intermittency factor. When $\Omega = .5$, $r/z = 0.16$, approximately.

The rms variation in the radial interface distance between the turbulent and nonturbulent zones is roughly four percent of the axial length (Corrsin and Kistler, 1954). This rms value gives an indication of the size of the eddies at the jet boundary. It is close to the lateral length scale determined by Laurence (1956).

Region III is the mixing zone that lies between the potential core and the undisturbed fluid. It is characterized by a high velocity gradient and high intensity of turbulence. In region IV, the mixing zone and potential core merge and form a totally turbulent pattern.

The essential features of a round jet can be easily understood with a plug flow model that assumes that the velocity in a jet is dependent only on the axial distance from the nozzle (independent of the radial position within the jet). With such a model, the jet boundary is well defined since the velocity is discontinuous across it. From visual observations, the total jet angle θ appears to be roughly fourteen degrees. Donald and Singer (1959) observed that jet angles increased with kinematic viscosity of the fluid. Such observations, although interesting, cannot be used quantitatively because the jet angle of interest is a momentum angle and not a visual angle indicating the jet boundary.

Assuming that the pressure is constant throughout the jet and the nozzle exit regions, we can relate conditions in the jet to those at the nozzle exit with the following equation:

$$w_j \overline{U}_j = w_n \overline{U}_n \tag{15}$$

where w_j and $\overline{U}_j$ are the mass flow rate and the jet velocity at a distance z from the nozzle (see Fig. 7). The subscript n refers to values in the throat of the nozzle. Diameter of the jet at z is given by

$$d_j = d_n + 2z \tan(\theta/2) \tag{16}$$

Each of the mass flow rates in Eq. (15) can be expressed as a function of velocity, density (ρ) and cross sectional area A:

$$w_n = \rho_n A_n \overline{U}_n \tag{17}$$

$$w_j = \rho_j A_j \overline{U}_j \tag{18}$$

An expression for velocity decay can be obtained by combining Eqs. (15) - (18):

$$\overline{U}_j / \overline{U}_n = \sqrt{\rho_n/\rho_j} \Big/ [1 + 2(z/d_n)\tan(\theta/2)] \tag{19}$$

The changes in mass flow rate can also be determined by this approach:

$$w_j/w_n = \sqrt{\rho_j/\rho_n}[1 + 2(z/d_n)\tan(\theta/2)] \tag{20}$$

when $\rho_j = \rho_n$, $\theta = 14$ degrees and z is large, Eq. (20) reduces to

$$w_j/w_n = 0.25(z/d_n) \tag{21}$$

This equation predicts that every four nozzle diameters the jet will entrain a mass flow rate equal to the nozzle flow rate.

Using eddy viscosity and Prandtl mixing length models, several researchers have developed theoretical estimates of jet phenomena. This work is summarized by Schlichting (1968) and by Harsha (1971). A simpler approach will be followed here to obtain an improvement over the plug flow model. This approach will recognize the radial variation in the axial velocity. Inside the potential core, the jet centerline velocity $\overline{U}_c$ remains constant and equal to $\overline{U}_n$. Beyond the potential core ($z > z_c$), $\overline{U}_c$ decreases with d_n/z. This behavior can be stated mathematically as Eqs. (22) and (23):

$$\overline{U}_c / \overline{U}_n = 1 \text{ at } z/d_n \leqslant z_c/d_n \tag{22}$$

$$\overline{U}_c / \overline{U}_n = (z_c/d_n)(d_n/z) \text{ at } z/d_n > z_c/d_n \tag{23}$$

Harsha has given the following expression for the core length in terms of Reynolds number:

$$z_c/d_n = 2.13\, N_{Re}^{0.097} \tag{24}$$

This equation is a good fit of the available data on air jets for Reynolds numbers between 10^4 and 10^5. Data of Hrycak *et al.* (1970) indicates that the dimensionless potential core length attains a maximum value of approximately 20 at a Reynolds number of 1,000, reduces to a minimum value of roughly 5.5 at a Reynolds number of 4,000, and levels out at about 6.5 for Reynolds numbers between 10,000 and 100,000.

The velocity profile at $z > z_c$ can be assumed to approximate a Gaussian distribution:

$$\overline{U}_z / \overline{U}_c = e^{-\psi r^2} \tag{25}$$

where the constant ψ is estimated from a momentum balance. With nozzle and jet fluids of the same density, the momentum and mass balances become

$$d_n^2\, \overline{U}_n^2 = 8\int_0^{\infty} \overline{U}_z^2 r dr \tag{26}$$

$$w_j/\rho = 2\pi\int_0^{\infty} \overline{U}_z r dr \tag{27}$$

From Eqs. (25) - (27), it is easily shown that

$$w_j/w_n = 2(d_n/z_c)(z/d_n) \tag{28}$$

Results for three different Reynolds numbers are presented in Table IV. These flow rate ratios could also be expressed as concentrations.

A plane jet can form from a slot or from a closely spaced row of orifices. Unlike circular jets, the Reynolds number of the jet structure increases with $\sqrt{z}$. The

TABLE IV *Circular Jet Characteristics*

N_{Re}	z_c/d_n	$w_j d_n/w_n z$
10^4	5.20	.384
10^5	6.51	.307
10^6	8.14	.246

dimensionless axial velocity and entrainment rate are given by a model reported by Schlichting (1968):

$$\overline{U}_z / \overline{U}_n = 2.40\sqrt{h/z}[1 - \tanh^2(7.67y/z)] \qquad (29)$$

$$w_{js}/w_{ns} = 0.625\sqrt{z/h} \qquad (30)$$

where h, y and w_{js} are the slot width, distance from the jet centerline and the mass rate in the jet per unit length of a slot, respectively.

As shown in Fig. 8, the radial wall jet is formed

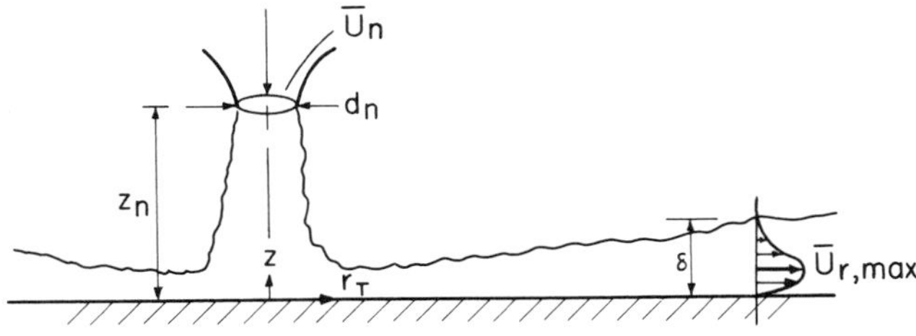

Fig. 8. Radial wall jet

by directing an axisymmetric jet flow perpendicular to a wall.

A number of analytical and experimental studies are available on radial wall jets; representative of these are the works of Hrycak *et al.* (1970), Dawson and Trass (1966) and Skifstad (1969). These and other studies give a fairly complete picture of the fluid dynamic behavior of radial wall jets.

A radial wall jet may consist of an axisymmetric free jet, a turning region and a wall jet region. If the r_T/z_n ratio is greater than approximately 0.5, the wall jet structure exhibits only minor dependence on the nozzle clearance z_n. The radial velocity goes through a maximum near the wall; it can be estimated by Eq. (31), given by Hrycak *et al.*:

$$U_{r,max}/\overline{U}_n = 1.4(dn/r_T)^{1.12} \tag{31}$$

For $r_T/z_n \geqslant 0.5$, Dawson and Trass (1966) found the overall height of the radial jet to be

$$\delta/z_n = 0.11(r_T/z_n)^{0.88} \tag{32}$$

As mentioned in an earlier section, many mixers are dependent on the behavior of circular jets in a cross stream. The designer needs to understand the jet penetration capability, the concentration decay characteristics and the effect of adjacent jets on penetration and concentration characteristics.

From the plug flow model, it is possible to develop a highly simplified penetration theory for a strong jet in a cross flow. At an axial distance z from the nozzle exit, the jet centerline is deflected a small distance s by the cross stream velocity $\overline{U}_s$. A short element of the jet Δz is accelerated by a drag force in accordance with Newton's second law:

$$C_D d_j \Delta z \rho_s \overline{U}_s^2/2 \simeq \frac{d}{dt} (\pi d_j^2/4)(\rho_j + \rho_s)\Delta z(ds/dt) \quad (33)$$

The right hand side of this equation contains the cross stream density ρ_s to account for the added mass of the jet. To manipulate the equation, the drag coefficient C_D will be assumed constant and ρ_j will be set equal to ρ_s. With $\overline{U}_j = dz/dt$ and Eqs. (16) and (19) simplified for large z, Eq. (33) can be integrated to give an equation for the penetration length z:

$$z/d_n \simeq \{18\pi/4C_D \tan(\theta/2)\}^{1/3}\{\sqrt{\rho_n/\rho_s}(\overline{U}_n/\overline{U}_s)\}^{2/3} \{s/d_n\}^{1/3} \quad (34)$$

If $C_D \simeq 1$, and $\theta \simeq 14^o$, then

$$z/d_n \simeq 4.9\{\sqrt{\rho_n/\rho_s}(\overline{U}_n/\overline{U}_s)\}^{2/3}\{s/d_n\}^{1/3} \quad (35)$$

Equation (35) can be compared with Eq. (36) given by Patrick (1967) and with Eq. (37) given by Callaghan and Ruggeri (1948)

$$z/d_n = 1.0\{\sqrt{\rho_n/\rho_s}(\overline{U}_n/\overline{U}_s)\}^{0.85}\{s/d_n\}^{0.38} \quad (36)$$

$$z/d_n = 1.91\{(\rho_n/\rho_s)(\overline{U}_n/\overline{U}_s)\}^{0.606}\{s/d_n\}^{0.303} \quad (37)$$

Although the form of Eq. (35) is okay, the value of 4.9 is roughly three times what it should be. Evidently, jet-cross stream interaction reduces jet penetration substantially.

Differences between the Patrick and Callaghan-Ruggeri equations are due in part to their different methods of determing the jet axis. For a given penetration distance z, Patrick searched for the position of the maximum velocity or concentration and defined that point as one on the jet axis. Using a 400°F jet, Callaghan and Ruggeri defined the jet axis or penetration position as the location where the temperature was 1°F above the cross stream temperature.

Patrick's expression for the decay jet centerline concentration is given here as Eq. (38):

$$1/\bar{A}_c = [(z_1/d_n)\exp\{7.8\sqrt{\rho_s/\rho_n}(\bar{U}_s/\bar{U}_n) - 1.856\}]^{1.18} \qquad (38)$$

where $\bar{A}_c$ and z_1 are the concentration of nozzle fluid at the jet centerline and the axial arc length in the jet, respectively. Eqs. (36) and (38) are valid for

$$0 < \sqrt{\rho_n/\rho_s}(\bar{U}_s/\bar{U}_n) \leq 0.152 \qquad (39)$$

Patrick reports that the $\bar{U}_s = 0$ form of Eq. (38) is in agreement with measurements of Hinze and van der Hegge Zijnen.

Multiple circular jets in a cross flow interact with each other in a complicated way. If the orifices are spaced too close together, the jets will feed on each other and inhibit mixing. Widely spaced orifices may allow the cross flowing stream to pass between the orifices, thereby bypassing the mixer. Walker and Kors (1973) made a thorough study of temperature mixing from multiple jet injection. They found that mixing and jet penetration increased with increasing momentum ratio $\rho_n\bar{U}_n^2/(\rho_s\bar{U}_s^2)$, with increasing orifice diameter and with increasing orifice pitch-to-diameter ratios. An orifice pitch-to-diameter ratio as low as two was found to have a deterimental effect on penetration and mixing. Although no optimum configuration was found by Walker and Kors, it is evident from their data that good mixing can be achieved with a pitch-to-diameter ratio of three or four, a mixing cell height-to-diameter ratio of roughly eight and a momentum ratio greater than six. The mixing cell height would be equal to the pipe radius for the types B and C mixers and to half the clearance between adjacent fingers for the type A mixer.

Multiple jets will have good penetration characteristics if they are designed (with Eq. 36) to deflect one to four times the mixing cell dimension when they have penetrated the mixing cell. Strongly penetrating jets will cause excessive pressure drop and will develop an upstream recirculation pattern within the mixing device. Weak jets do not mix satisfactorily.

Swirl enhances jet mixing (Fejer *et al.*1969). Velocity and concentration both decay more rapidly in a swirling jet than in a non-swirling one. Because of these phenomena and the toroidal recirculation pattern formed in the jet, swirl effectively shortens and stabilizes flames (Syred and Beér, 1974). It is commonly used in combusion processes.

III.B. *JET MIXING IN TANKS*

Because of their tendency to entrain large volumes of surrounding fluid, free jets are often used to mix the contents of large tanks. The mixing objective may be to combine two or more feed components, to suspend traces of solids or simply to blend the tank contents and thereby reduce the time variations of concentration in a flowing stream. The jet flow could come from a separate circulating pump or from the normal flow into the tank. Although the jet nozzle is usually convergent, a line size nozzle will also effect mixing.

Since specific design procedures for jet mixing systems are given elsewhere (The Oil and Gas Journal fourteen part series, September 6, 1954 to January 14, 1955), detailed design methods will not be given here; rather, this section will include selected topics on jet mixing phenomena. It is aimed at improving the range of applicability and performance of jet mixers.

Jet mixers generally consume more energy than mechanical agitation devices such as propeller mixers (Rushton, 1954). Considering the combined operating cost and equipment investment, they may be more or less expensive than propeller mixers. Some jet mixers utilize energy more efficiently than others. The designer can design a new system or modify an existing one to minimize energy consumption. Although such energy optimization may not lead to the most economical design, it will approach the optimum more closely than is possible with conventional design methods.

In some circumstances, it is necessary to jet mix a tank in a specified time. If the fluids in the tank are miscible and of low viscosity, the dimensionally consistent equation of Fossett and Prosser (1949) will lead to a good estimate of mixing time:

$$t = 8d_T^2 \Big/ \sqrt{Q_n \overline{U}_n} \tag{40}$$

t, d_T, Q_n and $\overline{U}_n$ are mixing time, tank diameter, nozzle volumetric flow rate and nozzle velocity, respectively. This and other nomenclature are shown in Fig. 9. Reynolds number

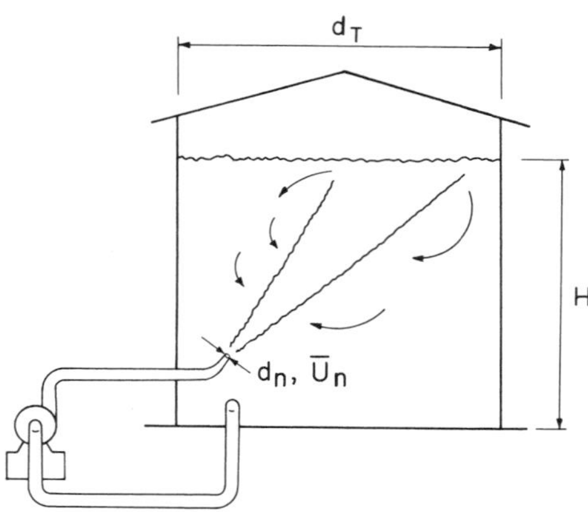

Fig. 9. Mixing tank

in the Fossett and Prosser tests was between 4,500 and 80,000. The constant of eight is applicable if the time taken to inject the second fluid is less than half the total mixing time.

Fox and Gex (1956) plotted a dimensionless parameter against Reynolds number:

$$\Gamma = t(d_n \overline{U}_n)^{2/3} g^{1/6} / H^{1/2} d_T \tag{41}$$

Although the Fox and Gex method covers a broader range of Reynolds numbers than the Fossett and Prosser correlation, the dependence on gravitational acceleration is suspect.

Gray (1966) summarizes the studies of Fossett and Prosser, Fox and Gex and Okita and Oyama, who obtained an equation that can be converted to a form similar to that of Eq. (40)

$$t = 4.6 d_T^{3/2} H^{1/2} \Big/ \sqrt{Q_n \overline{U}_n} \tag{42}$$

Since $d_n \overline{U}_n = \sqrt{4 Q_n \overline{U}_n / \pi}$, the Fox and Gex correlation can also be converted to the form of Eqs. (40) and (42). Consequently, if the mix time, tank diameter and liquid height are all specified, mixing is dependent on a momentum flux term M_n, where

$$M_n = Q_n \overline{U}_n \tag{43}$$

Even if there is a Reynolds number dependence, the added specification of kinematic viscosity will fix M_n, since

$$N_{Re} = \sqrt{4 M_n / \pi \nu^2} \tag{44}$$

In the next section, the momentum flux will be shown to be important in suspending solids.

Clearly, a particular mixing task can be performed in many different ways if M_n isn't changed. Large $\overline{U}_n$ and low Q_n will result in a high head, low volume pump.

The frictional losses in the Fig. 9 circulating system are the sum of those in the pipe and in the nozzle:

$$-\Delta p = (K_p \rho \overline{U}_p^2/2) + (K_n \rho \overline{U}_n^2/2) \tag{45}$$

where K_p, K_n, $\overline{U}_p$ and $\overline{U}_n$ are the velocity head loss through the piping, velocity head loss through the nozzle, average pipe velocity and nozzle velocity, respectively. The power expended, P, is

$$P = (K_p \rho \overline{U}_p^2 Q_n/2) + (K_n \rho \overline{U}_n^2 Q_n/2) \tag{46}$$

Substituting the momentum flux M_n gives

$$P = (K_p \rho \overline{U}_p^2 Q_n/2) + (K_n \rho M_n^2/2Q_n) \tag{47}$$

If a new circulating system (including pump) were being sized for fixed velocity $\overline{U}_p$ and M_n, it is easily shown from Eq. (47) that the minimum energy usage will occur when

$$Q_n = (M_n/\ \overline{U}_p)\sqrt{K_n/K_p} \tag{48}$$

If the piping is existing but the nozzle and the pump are being sized, $\overline{U}_p = Q_n/A_p$, and

$$Q_n = (M_n A_p K_n/3K_p)^{1/4} \tag{49}$$

Both Eqs. (48) and (49) assume that the pump will have an efficiency independent of the flow rate Q_n.

An existing jet mixing system can also be optimized to ensure the maximum momentum flux M_n. Typically, a centrifugal pump curve can be fit to an equation of the form:

$$F = F_0 - GQ_n^2 \tag{50}$$

where F is the pump head, F_0 is the head at shutoff and G is an empirical constant. Since the pressure drop head is equal to the pump head, Eq. (45) (expressed in terms of liquid head) can be equated to the right hand side of Eq. (50). When the resultant equation is differentiated and dM_n/dQ_n is set equal to zero, it is readily shown that the maximum momentum flux will be realized when

$$Q_n = [F_0/\{(K_p/gA_p^2) + 2G\}] \tag{51}$$

The designer may wish to optimize systems too complex to treat analytically. Such systems may be amenable to a graphical approach. The momentum term or power could be plotted against the flow rate, and the optimum rate determined from the shape of the curve.

III.C. *SOLIDS SUSPENSION WITH JET MIXERS*

When a liquid containing a trace of solids is pumped into a tank, the solids can settle and build up a layer in the bottom of the tank. Such build-up will eventually reduce the available tank volume. It can even cause a contamination problem if different fluids are stored in the tank. Cleaning can be very difficult, particularly if the solids harden and stick to each other.

Both side entering propeller mixers (Wilson, 1954) and vertical agitators have been used for solids suspension. Even though there are a few scale-up techniques available for the design of such mixers, the suspension mechanism is complex and scale-up problems are formidable (see Oldshue, 1969 and Gray, 1966, 1973).

A radial wall jet should be a highly effective device for suspending solids in a tank. It will create a horizontal velocity to move the solid particles and turbulence that will suspend them. An open pipe or nozzle can be centered on the tank axis and placed perhaps three nozzle diameters above the bottom of the tank. The jet, which is directed against the bottom of the tank, may erode the bottom unless an impingement plate is installed to protect the tank.

Insight into radial wall jet mixing of solids can be gained from studies of deposition in horizontal piping, several of which are summarized by Govier and Aziz (1972). The work of Thomas (1961, 1962 and 1964) is of particular value because it can be extended to non-pipe flows. To derive expressions for the minimum transport conditions, Thomas divided the flow regime into two parts, depending on whether the particles were larger than or smaller than the viscous sub-layer thickness. Expressed in terms of the friction velocity u^*, this sub-layer is taken to extend a distance δ_L from the wall:

$$\delta_L = 5\nu/u^* \tag{52}$$

When the particle diameter d_p is less than δ_L, Thomas found that the condition for minimum transport was

$$U_t/u^*_m = 0.010(d_p u^*_m/\nu)^{2.71} \tag{53}$$

From Eqs. (52) and (53), it is easily shown that Eq. (53) is valid when the particle Reynolds number $N_{Re,p} = d_p U_t/\nu < 4$. The friction velocity can be estimated from

$$u^* = \overline{U}_{r,max}\sqrt{f/2} \tag{54}$$

With the Fanning friction factor $f = 0.005$ and the radial dimension $r_T = d_T/2$, Eq. (31) can be combined with Eqs. (53) and (54) to obtain an expression for the required momentum flux:

$$M_n = 406U_t^2 d_T (d_T/d_n)^{0.24} (\nu/d_p U_t)^{1.46} \tag{55}$$

For particle diameters larger than the viscous sub-layer thickness, Thomas (1964) formulated an analysis in terms of turbulence properties. He gave the condition for minimum transport as

$$\varepsilon^{2/3} = g d_p^{1/3} (\rho_p - \rho)/\rho \qquad (56)$$

Although the rate of energy dissipation per unit mass would be expected to vary drastically over the δ dimension of the jet, a rough approximation for ε is given by Eq. (57):*

$$\varepsilon \simeq f\overline{U}_{r,max}/\delta \qquad (57)$$

Particles larger than the viscous sub-layer thickness will tend to be in the intermediate region where the drag coefficient is approximately (Boucher and Alves, 1973)

$$C_D = 18.5/N_{Re,p}^{0.60} \qquad (58)$$

The particle settling velocity is

$$U_t = \{4 g d_p (\rho_p - \rho)/3\rho C_D\}^{1/2} \qquad (59)$$

With $f = 0.005$ and $r_T = d_T/2$, Eqs. (31) and (32) can be combined with Eqs. (56) - (59) to give

$$M_n = 6 U_t^2 d_T^2 (\nu/d_p U_t)^{0.60} (d_T/d_p)^{2/3} (d_T/d_n)^{0.24} (z_n/d_T)^{0.08} \qquad (60)$$

In his work with pipe flow, Thomas (1964) obtained good agreement with liquid suspension data using Eq. (56). A correction that he suggested would reduce the Eq. (60) coefficient to 2.5.

An even cruder approach to this problem is useful

---

*For pipe flow, the data of Laufer (1954) indicate that the average rate of turbulent energy dissipation over the pipe cross section is approximately $\varepsilon = f\overline{U}_p^3/d_0$, which is one half of the total energy available from pressure drop. The remaining half undergoes direct viscous dissipation near the wall. The local turbulent dissipation rate determined by Laufer can be estimated with surprising accuracy from the following equation (Landau and Lifshitz, 1959): $\varepsilon = 2.4u^{*3}/y$, where y is the distance from the pipe wall. The equation is accurate from the centerline to the viscous sublayer. It can be compared to Eq. (8) given earlier.

when particle diameters and fluid viscosities are ill defined. The required friction velocity for minimum transport is taken as ω times the particle settling velocity. A combination of Eqs. (31) and (54), when expressed in terms of M_n gives

$$M_n = 34\omega^2 U_t^2 d_T^2 (d_T/d_n)^{0.24} \tag{61}$$

The value of ω should be between one and four in most applications.

Although Eqs. (55), (60) and (61) are rough approximations that have not been thoroughly tested, they do provide an insight into the suspension mechanism that is useful for scale-up and for preliminary design work. It is interesting to note that all three equations show that the momentum flux term M_n is of primary importance in mixer design; hence, the optimization techniques discussed in the previous section also apply to solids suspension. It is necessary to assume a value of the nozzle diameter d_n to use them.

IV. MISCELLANEOUS TURBULENCE THEORY APPLICATIONS

IV.A. *STRATIFICATION IN HORIZONTAL TUBULAR REACTORS*

In the chemical industry, tubular reactors are commonly used to produce liquid chemicals. When a reaction is conducted adiabatically, economic pressures can force the designer to use a configuration that may cause stratification or other non-plug flow behavior.

From data obtained in a pilot plant reactor or other small reactor, the designer usually knows the required residence time or reactor volume. He must decide what length and diameter will give him satisfactory plug flow behavior. He could use a long, small diameter reactor to ensure tubularity, but it would be expensive and might cause layout problems. Heat losses and pressure drop could also be excessive.

From an axial dispersion calculation, the designer might conclude that a reactor with a fifty to one length-to-diameter ratio would exhibit satisfactory plug flow behavior. With such a small length-to-diameter ratio, the reactor entrance piping would be smaller than the reactor itself, and he must take care to ensure that liquid enters the reactor in a way that will avoid a gross distortion of the velocity profile. Devices that will minimize disturbances include diffusers, perforated plates, distribution pipes and packing.

Convergence effects at the outlet have little effect on the reactor tubularity and can be neglected. It can be shown from an inviscid flow analysis that a fluid approaching a converging outlet must be within approximately three-fourths of the reactor diameter before its plug-flow velocity profile is altered appreciably.

In reactors of this type, stratification may cause problems with reactor operation (Simpson, 1974). Even though all reactants and products may be totally miscible and the Reynolds number may be well into the turbulent regime, fluids of different compositions and temperatures can layer. This phenomena can reduce reaction efficiency significantly and, if the temperature gradient is large enough, it may even cause bowing of the reactor.

Both the cause and cure of stratification can be found in turbulence effects. The axial velocity in a pipe is not uniform, even in turbulent flow. Fluid near the wall moves more slowly than that near the center; consequently, it will have achieved a higher conversion at a given cross section of the reactor. If the reaction has a significant heat of reaction, the wall liquid temperature will be different from that of the bulk liquid, and the higher density material will migrate to the bottom. This phenomena will occur even if the reactor is well insulated.

Based on an order of magnitude estimate for eddy size and eddy velocity, Simpson (1974) presented a technique for avoiding stratification. He showed that stratification could be eliminated if a reactor is designed so that the absolute value of the density difference between reacting species is less than that predicted by Eq. (62):

$$|\Delta\rho_c| \simeq \rho\overline{U}_p^2/2gd_0^2 \qquad (62)$$

In this equation, $\Delta\rho_c$, ρ, $\overline{U}_p$, g and d_0 are the critical density difference, density of one of the species, average pipe velocity, acceleration of gravity and pipe diameter, respectively. This equation has been found to be in agreement with a few observations on operating reactors. The absolute value of the reactant-less-product densities was compared with the Eq. (62) estimate.

Although not directly applicable to reactors, mixing studies have been conducted with stratified salt water and fresh water solutions. Based on Turner's analysis (1973), a pipe would be considered well mixed if the right hand side of Eq. (62) were multiplied by .13 and stratified if multiplied by 1.3. These multipliers may be somewhat larger in a reactor because stratification would develop gradually.

IV.B. *ATMOSPHERIC DIFFUSION*

In recent years, industrial environmental specialists and safety experts have become increasingly concerned with the mixing phenomenon called atmospheric diffusion. To provide a safe and healthy environment for plant workers and for the local community, they must understand how to assess the impact of accidental chemical spills even though the incident probability is very low. In addition to a knowledge of the available correlations, they need an insight into the nature of turbulence itself.

Turbulence theory has played an important role in the development of diffusion technology. In their pioneering work on atmospheric diffusion, Bosanquet and Pearson (1936) recognized the importance of turbulence phenomena. Pasquill (1962) devoted about half of his classical work on diffusion to fundamental turbulence behavior. Turbulence was also emphasized in the recent book by Seinfeld (1975).

There are many gaps in diffusion technology. The effect of obstacles on diffusion cannot be predicted quantitatively, nor are the diffusive characteristics of flashing streams well understood. Although the solutions to these and other diffusion problems will require experimental studies, the application of turbulence theory will minimize the experimentation and will provide a sound basis for empirical correlation of the results.

IV.C. *PREDICTION OF BUBBLE AND DROP SIZE*

Trying to understand drop or bubble transport processes can be a frustrating experience. Before any meaningful calculations can be undertaken, it is usually necessary to estimate particle size or particle size distribution. Often, the calculation technique is very sensitive to particle size, yet the size estimates are usually very crude. Moreover, different methods may give contradictory results.

There are a wide variety of particulate transport processes. They include the spray quenching of gases, oil burning, contacting of two immiscible liquids and pipeline or tank type bubble contacting.

To understand a given transport process without a good particle size estimate, one must either use order of magnitude engineering or undertake an experimental effort. As distasteful as a crude estimating technique may be, it may be preferred because of the time delay, costs or scale-up problems associated with an experimental program. Rough estimating methods are also useful to those engaged in trouble-shooting activities and to those charged with developing scale-up methods for plant design.

When sprays are produced by hydraulic atomizers, the drop size can sometimes be estimated from manufacturers' literature. Another estimating technique for spray drop size involved the use of a critical Weber number (Hinze, 1955). A drop exposed to an increasing relative gas velocity will experience an increasing imbalance in the aerodynamic-to-surface-force ratio. At a critical value of the Weber number, breakup occurs.

Hinze argued that an emulsion of a non-coalescing liquid-liquid mixture in turbulent motion is also subject to Weber number break-up, but that the Weber number should be based on the rms velocity fluctuations instead of a relative velocity. After obtaining an expression for this velocity fluctuation, he showed that the maximum particle diameter is given by

$$d_{p,max}(\rho_c/\sigma)^{3/5}\varepsilon^{2/5} = C \tag{63}$$

where $d_{p,max}$, ρ_c, σ and C are maximum drop size, density of continuous phase, interfacial tension and a constant. From reported data on coaxial cylinders (inner cylinder rotating), Hinze determined that C = .725 if $d_{p,max}$ is taken as the size that, together with all smaller sizes, contains ninety-five volume percent of the drops.

Calderbank (1958) obtained correlations similar to Eq. (63) from his work with gas-liquid and liquid-liquid mixtures in mechanically agitated vessels. Sevik and Park (1973) found that C = 1.15 for bubbles in a water jet and Miller

(1974) reported that C = 1.93 for the stable bubble diameter in a gas sparged tank.

The energy dissipation per unit mass ε can be estimated for certain situations. Miller (1974) explained how to estimate dissipation in sparged and mechanically agitated systems. Hughmark (1968) presented energy dissipation correlations for orifices and venturis. Eq. (63) can be applied to pipe flow since ε can be estimated from $\varepsilon = f\bar{U}_p^3 d_0$. It is less obvious how ε should be estimated in other situations, such as with immiscible liquids flowing through a pump.

Sleicher (1962) argued that Eq. (63) should not be applied to pipe flow. His immiscible liquid data clearly showed that drops break up very near a wall where the isotropic turbulence assumption is least applicable. Even though wall break-up is expected from turbulence considerations, since the energy dissipation rate near the laminar sublayer may be ten times the average value, Sleicher's criticism is valid. The non-isotropic turbulence found in essentially all chemical processing equipment precludes any rigorous application of Eq. (63). Nevertheless, the equation and modifications of it are of utility to those who would flounder on drop process calculations without such techniques.

Prominent in the field of drop and bubble processes is the monograph of V.G. Levich (1962) which contains numerous theoretical analyses of break-up phenomena. Among them are theories of drop break-up in piping that discern between bulk and wall mechanisms. Despite the fact that it contains mostly rough approximations that make the reader wary and reluctant to apply the work, Levich's book contains an insight that is rare in this field.

V. REFERENCES

Batchelor, G.K. (1967) AN INTRODUCTION TO FLUID DYNAMICS, Cambridge University Press, Cambridge.

Beek, J.,Jr. and Miller, R.S. (1959) *Chem. Eng. Prog. Symposium ser. No. 25. 55,* 23.

Beér, J.M. and Chigier, N.A. (1972) COMBUSTION AERODYNAMICS, Halstead Press Division, John Wiley & Sons, Inc., New York.

Boucher, D.F. and Lapple, C.E. (1948) *Chem. Eng. Prog. 44(2),* 117.

Boucher, D.F. and Alves, G.E. (1973) *In* CHEMICAL ENGINEERS HANDBOOK (R. H. Perry and C. H. Chilton, eds.), 5 - 1, McGraw-Hill, New York.

Bosanquet, C.H. and Pearson, J.L. (1936) *Trans. Faraday Soc. 36,* 1249.

Brodkey, R.S. (1966) *A.I.Ch.E.J. 12,* 403.

Brodkey, R.S. (1967) THE PHENOMENA OF FLUID MOTIONS, Addison-Wesley, Reading, Massachusetts.

Calderbank, P.H. (1958) *Trans. Inst. Chem. Eng. 36,* 443.

Callaghan, E.E. and Ruggeri, R.S. (1948) NACA Technical Note 1615.

Chilton, T.H. and Genereaux, R.P. (1930) *A.I.Ch.E. TRANS. 25,* 102.

Corrsin, S. and Kistler, A.L. (1954) NACA Tech. Note 3133.

Corrsin, S. (1957) *A.I.Ch.E.J. 3,* 329.

Corrsin, S. (1964) *A.I.Ch.E.J. 10,* 870.

Dawson, D.A. and Trass, O. (1966) *Can. J. Chem. Eng. 44,* 121.

Donald, M.B. and Singer, H. (1959) *Trans. Inst. Chem. Eng. 37,* 255.

Fejer, A.A., Hermann, W.G. and Torda, T.P. (1969) Aerospace Research Laboratories Report ARL 69-0175.

Fossett, H. and Prosser, L.E. (1949) *J. Inst. Mech. Eng. 160,* 224.

Fox, E.A. and Gex, V.E. (1956)*A.I.Ch.E.J. 2,* 539.

Friehe, C.A., VanAtta, C.W. and Gibson, C.H. (1972) AGARD Conference Proceedings No. 93.

Gegner, J.P. and Brodkey, R.S. (1966) *A.I.Ch.E.J. 12,* 817.

Govier, G.W. and Aziz, K. (1972) THE FLOW OF COMPLEX MIXTURES IN PIPES, Van Nostrand Reinhold, New York.

Gray, J.B. (1966) *In* MIXING-THEORY AND PRACTICE (V. W. Uhl and J. B. Gray, eds.) Academic Press, New York.

Gray, J.B. (1973) *In* CHEMICAL ENGINEERS' HANDBOOK (R. H. Perry and C. H. Chilton, eds.), 19-3, McGraw-Hill, New York.

Harsha, P.T. (1971) Arnold Engineering Development Center, Technical Report AEDC-TR-71-36.

Hartung, K.H. and Hiby, J.W. (1972) *Chem. Ing. Tech. 44,* 1051.

Hiby, J.W. (1971) *Chem. Ing. Tech. 43,* 378.

Hinze, J.O. (1955) *A.I.Ch.E.J. 1,* 289.

Hrycak, P., Lee, D.T., Gauntner, J.W. and Livingood, J.N.B. (1970) NASA Technical Note D-5690.

Hughes, R.R. (1957) *Ind. Eng. Chem. 49,* 947.

Hughmark, G.A. (1968) Paper presented at the sixty-first annual meeting of the A.I.Ch.E., Los Angeles, California.

Ito, H. (1960) *Trans. A.S.M.E., Ser. D. 82,* 131.

Jacobs, L.J.,Jr. (1973) Paper presented at the Engineering Foundation Research Conference, South Berwick, Maine.

Landau, L.D. and Lifshitz, E.M. (1959) FLUID MECHANICS, Pergamon Press, London.

Laufer, J. (1954) NACA Report 1174.

Laurence, J.C. (1956) NACA Report 1292.

Lee, J. and Brodkey, R.S. (1964) *A.I.Ch.E.J. 10,* 187.

Levenspiel, O. and Bischoff, K.B. (1963) *In* ADVANCES IN CHEMICAL ENGINEERING (T. B. Drew, J. W. Hoopes, Jr. and T. Vermeulen, eds.), *IV*, 95, Academic Press, New York.

Levich, V.G. (1962) PHYSICOCHEMICAL HYDRODYNAMICS, Prentice-Hall, Englewood Cliffs, New Jersey.

Miller, D.N. (1974) *A.I.Ch.E.J. 20,* 445.

Narayan, B.C. (1971) M. S. Thesis, Univerity of Tulsa.

Oldshue, J.Y. (1969) *Ind. Eng. Chem. 61(9)*, 79.

Pasquill, F. (1962) ATMOSPHERIC DIFFUSION, D. Van Nostrand, New York.

Patrick, M.A. (1967) *Trans. Inst. Chem. Eng. 45,* T16.

Reed, R.D. (1965), U. S. Patent No. 3,332,442.

Reed, R.D. (1974) Private Communication.

Rushton, J.H. (1954) *Pet. Refiner. 33(8)*, 101.

Schlichting, H. (1968) BOUNDARY-LAYER THEORY, 6th ed. McGraw-Hill, New York.

Seinfeld, J.H. (1975) AIR POLLUTION, McGraw-Hill, New York.

Sevik, M. and Park, S.H. (1973) *Trans. A.S.M.E.,J. of Fluids Eng. 95*, 53.

Simpson, L.L. (1968) *Chem. Eng. 75(13)*, 192.

Simpson, L.L. (1974) *Chem. Eng. Prog. 70(10)*, 77.

Singh, M. and Toor, H.L. (1974) *A.I.Ch.E.J. 20*, 1224.

Skifstad, J.G. (1969) Purdue University Jet Propulsion Center Technical Report AFAPL-TR-69-28.

Sleicher, C.A. (1962) *A.I.Ch.E.J. 8*, 471.

Son, J.S. (1972) U. S. Patent No. 3,702,619.

Syred, N. and Beer, J.M. (1974) *Combustion and Flame 23*, 143.

Thomas, D.G. (1961) *A.I.Ch.E.J. 7*, 423.

Thomas, D.G. (1962) *A.I.Ch.E.J. 8*, 373.

Thomas, D.G. (1964) *A.I.Ch.E.J. 10*, 303.

Townsend, A.A. (1961) *In* HANDBOOK OF FLUID DYNAMICS (V. L. Streeter, ed.) 10, McGraw-Hill, New York.

Treybal, R.E. (1963) LIQUID EXTRACTION, 2nd ed. McGraw-Hill, New York.

Turner, J.S. (1973) BUOYANCY EFFECTS IN FLUIDS, Cambridge University Press, Cambridge.

Walker, R.E. and Kors, D.L. (1973) NASA Report No. CR-121217.

Ward-Smith, A.J. (1971) PRESSURE LOSSES IN DUCTED FLOWS, Butterworths, London.

Wilson, N.G. (1954) *Oil Gas J. 53*, 165.

Woodhead, J.R., White, E.T. and Yesberg, D. (1971) *Can. J. Chem. Eng. 49*, 695.

Author Index

Numbers in italics refer to pages on which references can be found.

D

E

F

G

H

Y

Z

Subject Index

A 5
B 6
C 7
D 8
E 9
F 0
G 1
H 2
I 3
J 4